French Historical Studies

Volume 48 · *Number 2* · *May 2025*

Science, Technology, Mobility, and Modernity in French and Francophone Histories
APRIL G. SHELFORD and PETER S. SOPPELSA, Special Issue Editors

Introduction to the Special Issue

APRIL G. SHELFORD and PETER S. SOPPELSA

ABSTRACT   This special issue demonstrates how investigating topics related to the histories of science and technology illuminates themes of enduring importance in French and Franco-phone historical study—from policing female bodies to expressing patriotic anxieties. By documenting the circulation of people, ideas, information, artifacts, and technology within and beyond the Hexagon, these six articles help rethink French science and technology since the 1700s by problematizing the often-assumed links among science, technology, mobility, and modernity. Instead of seeing moments of change in science or technology as self-evident moments of modernity, the authors show how such moments unleash crosscurrents that pit modernizing projects against moves that resist modernization. Rather than any linear rela-tionship between science and technology, on the one hand, and modernity, on the other, these articles stress the consistent, contingent, and contested entanglement of these terms. Reflecting the history of science's shift away from great men and often state-sponsored sci-ences, they emphasize applied, "minor," and popular sciences practiced by individuals beyond the academically trained. Here, users and consumers of science and technology are often as important as its producers, and observers outside the magic circle of professionals debate the stakes in and value of innovations—from decrying moral perils to legitimating imperial projects.

KEYWORDS   science, technology, modernity, mobility, France

A French female painter disturbed by the human viscera in a waxwork Venus on display in an Italian museum . . . a master taxidermist packing up his natural history collection for shipment to Scotland . . . the skeptical greeting of British gaslight technology in Restoration Paris . . . an exiled Communard enclosing a vocabulary list of an Indigenous people in a letter home . . . well-heeled tourists motoring through French North Africa . . . a female pedestrian experiencing an urgent call of nature in Paris . . .

What do these vignettes, evoked by the articles in this issue, have in common? Quite a lot, it turns out. First, they demonstrate how investigating topics related to the histories of science and technology illuminates themes of enduring importance in French and Francophone historical study, from policing female bodies to expressing patriotic anxieties. Documenting the ceaseless circulation of people, ideas, information, artifacts, and technology within and beyond the

Hexagon, they do more than help us rethink French science and technology since the 1700s: they problematize the often-assumed links among science, technology, mobility, and modernity. Instead of seeing moments of change in science or technology as self-evident moments of modernity, they show how such change unleashes crosscurrents that pit modernizing projects against moves to contest and resist modernization. Therefore, rather than any linear relationship between science and technology, on the one hand, and modernity, on the other, this issue stresses the consistent, contingent, and contested entanglement of these terms. Reflecting the history of science's shift away from great men and major, often state-sponsored sciences, these articles emphasize applied, "minor," and popular sciences practiced by individuals beyond the academically trained. As such, they disrupt easy distinctions between mental and manual labor, the head and the hand. In these articles, users and consumers of science and technology are often as important as its producers, and observers outside the magic circle of professionals and experts debate the stakes in and value of innovations, from decrying moral perils, often centered on gender, to legitimating imperial projects.

In short, these six articles demonstrate how science and technology have been woven into the fabric of French society and culture. The remainder of this introduction highlights some of the main themes that thread through the issue, intertwining with one another and with other scholarly concerns. Suggestive rather than exhaustive, this introduction is also an appreciation of the conceptual richness of the whole created by these six articles and an invitation to scholars of French and Francophone histories to pursue such concepts in their own work.

Mobilities: National and Imperial Trajectories

Several articles here test the connections that scholars have often assumed exist among science, technology, modernity, and mobility by exploring whether those connections persist when entangled with national identity and empire. Historians and social scientists have often claimed that mobility is both a defining feature and a generator of modernity. Whether we call it globalization, David Harvey's "compression" of time and space, Wolfgang Schivelbusch's "industrialization" of time and space, or the nineteenth-century term *annihilation* of time and space, mobility signals an accelerating, shrinking, and interconnected world.[1] At the same time, science and technology have also been proposed as defining aspects of modernity. What happens when we put these two arguments together? On the one hand, mobility appears as a necessary support

1. Creswell, *On the Move*; Harvey, *Condition of Postmodernity*; Schivelbusch, *Railway Journey*.

for developing science and technology. The practices that make modern science and technology presume a highly mobile world, fit for a wide-ranging transit of people, specimens, data, and ideas. On the other hand, mobility can be seen as a consequence of science and technology, an outcome of powerful ideas, practices, and instruments that accelerate and disenchant the modern world.

Taking different forms, mobility literally generated the author's subject in each of these articles. Yotam A. Tsal shows in "Exporting French Nature" how Louis Dufresne's extensive natural history collection would not even have existed without the long-standing, continuously ramifying networks that brought specimens and scientific information into Paris from around the globe.[2] In Elizabeth Della Zazzera's article "The Illumination of Restoration Paris," a German inventor, Frederick Albert Winsor, brought the gas lamps he had pioneered in London to Paris, where they would enhance the mobility, safety, and surveillance of inhabitants at night. Years later in "Louise Michel et les savoirs de l'exil," a Communard exiled to New Caledonia paradoxically found in constraint "the rest required to think, to feel alive, to read, to write, and 'to exist a bit as a free being.'" Volny Fages, Jérôme Lamy, and Florian Mathieu show how she used intellectual and revolutionary epistolary networks to send news of her botanical, ethnographic, and pedagogical endeavors home. In the Haussmannized Paris of Louise Thiroux's "Les toilettes pour dames s'emparent du macadam," women traverse new boulevards that promoted pedestrian and other forms of mobility while still lacking in 1870 public facilities for relieving themselves that men had enjoyed since the July Monarchy. In "Automobiles, Entrepreneurs, and Empire," Zohar Sapir Dvir writes that "where France expanded its colonial presence, French tourists followed"—and they preferred to do so in automobiles, which required the creation of physical and mental infrastructure, such as roads and travel guides.

As "modernity" exists in relation to a "past" from which it has allegedly progressed or decisively broken, it inevitably acquires meanings and prompts value judgments that at first blush appear quite foreign to the brute facts, say, of a preserved bird or a carburetor. Thus the sale of Dufresne's collection to the University of Edinburgh was both material as embodied in seashells, bugs, and other natural specimens, and conceptual in Dufresne's proposed layout for its new home and his long-definitive manual on taxidermy. While the collection asserted the global empire of French science, it also became a source of national

---

2. On museums, scientific networks, and collecting in the early modern period, see Findlen, *Possessing Nature*; Findlen, *Empires of Knowledge*; Delbourgo, *Collecting the World*; MacClellan and Regourd, *Colonial Machine*; Bleichmar and Mancall, *Collecting Across Cultures*; and Schnapper, *Collections et collectionneurs dans la France du XVIIe siècle*.

pride for the Scottish institution that purchased it. A revolutionary feminist on her way to anarchism, Michel obviously found much to reject in the modern republican empire that consigned her quite literally to its margins. Nevertheless, although she empathized with the situation of the Kanaks among whom she lived and sought to change European perceptions of them as cannibals, she remained captive of a stadial view of human history that dovetailed with, even facilitated, the oppression of non-Europeans. Let us imagine, too, the car packed with French tourists evoked by Sapir Dvir's research, speeding past Roman ruins in North Africa. Such a scene was potentially edifying to the tourists on two counts: first, the French resumption of an ancient imperial project; second, their surpassing of it.

Yet Della Zazzera's article on street lighting in Paris shows how claims to modernity can be challenged. A period dubbed "Restoration" by people living through it was bound to have a complex relationship with any technology identified as "modern." Thus, people on the political Right and Left expressed ambivalence toward or simply rejected gas lights, which were already suspect because of their British origins. Romantic critics, who were also suspect because of their foreign inspiration, could condemn gas lights as merely "new," thus unnecessary and unworthy of a truly modern, "restored" French monarchy. And while gaslight would make navigating the city at night easier, it would also facilitate the surveillance of its increasingly mobile population. The gaslight debates thus reproduce a familiar ambivalence about technological change that gripped the industrial age as contemporaries juggled enthusiasm for and fear of modernizing technologies.[3]

Thiroux shows how ambivalence toward modernity was literally built into the architecture of the first "chalets" that served as public lavatories for women.[4] On the one hand, gas lights had now become synonymous with modernity, and they illuminated comfort stations whose gleaming interiors of marble and porcelain implemented the dictates of modern hygienic ideas. Yet the design of the first such structures was an urbanophobic "engineering fiction," evoking "the alleged healthful virtues of the mountains," according to Thiroux; in our view, these "Swiss" chalets also brought the nineteenth-century altitude cure to Paris, albeit with a surprising and somewhat scatological twist.[5]

In short, these articles confirm that modernity and technology form a "compelling tangle"[6] while demonstrating how that entanglement eschews tidy,

---

3. Rieger, *Technology and the Culture of Modernity.*
4. Ross, "Dirty Desire."
5. Jennings, *Imperial Heights.*
6. Misa et al., *Modernity and Technology.*

predictable relationships. They confirm, too, that modern France was a high-stakes battleground for modernity because of the ideological diversity of French political culture, as Della Zazzera shows.[7] In such contests, science and technology made modernity not more secure but more embattled.

Gender and Bodies: Moral Panics and Pedagogies

Thiroux's article and Margaret Carlyle's "Women and Anatomy Education in Enlightenment France" also encourage us to think about how technological innovations centering on the female body—its "nature" and its alleged capacity to provoke disorder—made the museum and the public lavatory places of moral peril despite their explicitly stated purposes of edification and education, on the one hand, and relief in safety and privacy, on the other.

Scholarship exploring the nexus of gender, women, and medicine in the early modern period is well established and very rich.[8] During the eighteenth century some individuals had more generous views of female capacities than Thiroux's medical writers in the nineteenth. Nevertheless, erroneous beliefs about the "nature" of the female body—hence the "nature" of women—persisted even as the Galenic synthesis gave way to new medical views. "Progressive" pedagogues such as Mme. de Genlis believed that girls should have knowledge of their own bodies as they transitioned into their social roles of marriage and child rearing—and some philosophers, such as Denis Diderot, agreed. An especially powerful way to communicate that knowledge was through closely monitored and carefully scaffolded encounters with detailed anatomical waxwork models of the female body, such as those constructed by Marie-Marguerite Biheron. Displays featuring these waxworks figured in other efforts to engage a wider public in science, such as lecture-demonstrations, though sometimes the perceived sensationalism of these events undercut their serious objectives. According to Carlyle, the finely crafted models meant to "bring women and otherwise genteel audiences into the fold of human anatomy without offending their polite sensibilities" could be confused with more "pedestrian" versions, that is, "the late-century fairground knock-offs or the models displayed in the 'salons de cire' found in the entertainment districts of cosmopolitan cities like Paris."[9] This fact, combined with the abundant production of libertine literature, aroused the suspicions of

---

7. Fox, *Savant and the State*; Levitt, "Science and Technology Beyond the Barricades."

8. Schiebinger, *Mind Has No Sex?*; Laqueur, *Making Sex*; Jordanova, *Sexual Visions*; Wilson, *Women and Medicine in the French Enlightenment*; Doig and Sturzer, *Women, Gender, and Disease*; Gelbart, *Minerva's French Sisters*; Vila, *Enlightenment and Pathology*.

9. For more on public presentations of science, Sutton, *Science for a Polite Society*; Lynn, *Popular Science and Public Opinion*; Lachapelle, *Conjuring Science*.

more conservatively minded individuals. They "viewed [such displays] as a cover for immorality and an easy outlet for debauchery"—suspicions that, of course, the notorious marquis de Sade was happy to confirm. In other words, if we posed the question "Can women be educated into the 'facts' of sexuality without being corrupted?" to Biheron's contemporaries, many would have responded with a resounding "No!"

Public lavatories for women were meant to serve the needs of the female *corps pissant*, but they also figured in the Haussmannization of Paris and the implementation of new imperatives of public hygiene. The latter could be perceived as particularly urgent given widespread belief in the "degeneration" of the nation; it also resulted in the oppression of social groups considered undesirable, such as prostitutes, to contain the spread of venereal disease.[10] Yet there could be more benign effects—if for the wrong reasons. Thiroux documents how medical theorists who maintained that the female body was "le lieu des excès, des hystéries, des maladies," and other problems also worried about "la bonne santé des vessies et plus généralement des organes reproducteurs." In other words, for the sake of their health, women should simply not have to hold it in. The creation of *pissoirs* for men had similar medical justifications, as Andrew Israel Ross shows.[11] His work also suggests that the city fathers, when initiating the female comfort stations, could hardly have been surprised that containing bacterial contagion was easier than containing social contagion—especially its challenges to bourgeois values. As *pissoirs* had become a site for homosexual encounters, so it was feared that women's facilities would become a theater of homosexuality, prostitution, theft, murder, and suicide. Enter the *gardienne* of a new style of chalet, who was charged with surveilling the clientele and reinforcing the moral standards, which the structures were intended to instill in their users.[12]

In short, these artifacts—anatomical models and toilets—prompted social concerns and agendas, often expressed in moralizing terms. In the world Carlyle describes, the young woman tutored properly in her own anatomy would be better equipped for her socially appropriate, morally upright role as a wife, ready to assume her conjugal and child-bearing duties—or, corrupted, she would become licentious, an enemy of her own happiness (correctly understood) and a source of social disorder. In Thiroux's Paris, the impetus for women's public facilities stemmed from a larger program to manage waste, thus making it a healthier place. Yet because personal cleanliness was considered a measure of bourgeois

10. Murad and Zylberman, *L'hygiène dans la République*; Quinlan, *Great Nation in Decline*; Latour, *Pasteurization of France*.
11. Ross, "Dirty Desire."
12. Both tasks were facilitated by the ubiquity of gaslights after the Second Empire's massive expansion of the network. See Clayson, *Illuminated Paris*.

respectability, comfort stations equipped with an increasing number of conveniences beyond the toilet mimicked the interior of bourgeois homes and architecturally tutored lower-class (hence socially deficient) females in the actions of a proper woman. But this "improving" social agenda was undermined when the public lavatory became a potential site for salacious, rather than clean, edifying behaviors. In sum, artifacts designed to educate about the "nature" and needs of the female body were believed to have the power to cue or modify certain behaviors in morally desirable or undesirable ways.

Applied and Popular Sciences

As history of science has moved away from the history of great men, it has simultaneously moved away from the assumption that professional scientists are the only (important) actors in the scientific past. Thus this special issue tells us almost nothing about famous scientists like Louis Pasteur and Marie Curie. While present here, doctors and engineers are sometimes antagonists rather than protagonists. Following currents in the social history of science and technology, these articles blur distinctions between mental and manual labor. They also hold that the users or consumers of science and technology are as worthy of scholarly consideration as its producers. If we accept that knowledge is made by diverse groups rather than by privileged individuals, then we raise significant questions about the boundaries of science. The articles in this issue explore these boundaries by examining applied and popular sciences, with the article on Louise Michel providing a nice review of historical literature on popular science.[13]

As we make these methodological moves, scholars must take care in translating the word *science* from French to English. In everyday terms, English science means the natural or physical sciences, while in French *science* means professional scholarship or knowledge making, a broader notion closer to the German *Wissenschaft*. Meanwhile, in scholarly terms, Anglophone science studies examine a range of natural and social sciences, while French science studies have traditionally focused on historicizing the natural or physical sciences. These varied and layered definitions of *science*/science help explain why recent historical research on the sciences has closely watched how historical actors defined and contested "science" while documenting how such definitions shift over time.[14]

---

13. See also Bertucci, *Artisanal Enlightenment*; Kim, *Imagined Empire*; Lachapelle, *Conjuring Science*; and Lachapelle, *Investigating the Supernatural*.

14. Latour and Bowker, "Booming Discipline Short of Discipline"; Freudenthal, "Historian of Science's Guide to France"; Freudenthal, "Science Studies in France"; Picard, "L'organisation de la science en France"; Pestre, "Pour une histoire sociale et culturelle des sciences"; Levin, "What the French Have to Say"; Fox, "Fashioning the Discipline"; Chartier, "Sciences et savoirs"; Simon, "Forum Retrospectives."

Historical actors and scholars alike have devoted close attention to the questions of which fields of inquiry are part of "science," who gets to participate in these fields, and with what authority. The articles by Carlyle, Tsal, Thiroux, and Fages et al. make important contributions to this conversation.[15]

Tsal and Fages et al. illustrate the recent scholarly trend of studying the applied sciences, deemphasizing "pure" science, such as the production of theories. These articles examine both descriptive and experimental methods in fields from botany, zoology, and taxidermy to ethnography and pedagogy as they relate them to social-historical questions about networking and gatekeeping. Who gets to participate in science: Taxidermists? Globetrotting specimen collectors? Exiled revolutionaries? Carlyle joins Fages et al. in tracing both the limits of women's and girls' participation in science and their efforts to access science in careful and creative ways. A related question for us is: Where and when do social networks, especially those that reach overseas, go from being legitimate and powerful, such as Dufresne's, to being illegitimate and marginalized, such as Michel's? Our contributors also show how professional science and popular science overlap consistently, but not always comfortably. The practice and practitioners of science, therefore, circulate not only among labs, fields, and scientific institutions but also in and out of museums, penal colonies, and people's homes. Michel was exiled for her participation in the Paris Commune, a punishment that was supposed to cut her off from French imperial social networks. But as the article demonstrates, her exile resulted not in disconnection but in different connections to France and to the world. We might also ask: Did she desire to maintain connections to the social networks behind mainstream science, or was she turning instead toward an alternative science, a revolutionary science, a subversive science? Finally, her case raises the question of how the social sciences change broader definitions of *science*.

Thiroux deals with hygiene as a stimulus to architectural and technical choices, as well as to social and moral ones. Hygiene was a prominent applied science in nineteenth-century France, as relevant in public toilets as it was in scientific academies or publications. Bruno Latour described hygiene as "a social movement of gigantic proportions that declared itself ready to take charge of everything."[16] But despite hygiene's towering ambitions to transform bodies, morals, society, and environments, it failed to establish itself through institutionalization and professionalization as a fully fledged scientific discipline. Rather,

15. Similar concerns about boundaries and definitions of science have also inspired a growing research field, the history of knowledge, which is designed to be more capacious and inclusive than the history of science. See Dupré and Somsen, "Forum."
16. Latour, *Pasteurization of France*, 33; see also Chevallier, *Le Paris moderne*.

it became what Lion Murad and Patrick Zylberman called a "thwarted utopia," which retreated into quotidian problems of cleaning bodies, behaviors, and bidets.[17] Because hygiene could not wrest its objects and methods of study away from competing fields of biology and medicine, by 1900 it was less a discipline of its own than an applied version of other sciences, a constellation of practices in bathing, grooming, housekeeping, medicine, and public health.[18]

Artifacts and Networks

In contrast to a special issue on mobility of *French Historical Studies* published in 2006, this issue shifts from state-directed mobility to freer-flowing movement, and it includes moving things alongside moving people.[19] It demonstrates how much mobility studies has evolved since its early days two decades ago, and how it has intersected with the scholarly interest in artifacts and networks and applied and popular sciences described above. The articles gathered here illustrate scholarship's increasing focus on concrete artifacts (material culture)—not only in histories of medicine and technology, where it might be expected, but also in the history of science, where artifacts are increasingly integrated into understandings of how knowledge is created. Teaching tools, visual aids, models, specimens, and instruments are strikingly mobile; they circulate through different contexts and different hands, in which they assume different meanings and consequences.[20]

The authors in this issue tell stories that connect the concrete artifacts of science and technology with the mobility of people and things: taxidermists and tourists, specimens, guidebooks, wax models, water, and gas. Carlyle shows how wax anatomical models changed symbolically and morally as they moved from boudoir to parlor, museum, university, or hospital. Tsal details a global transit in specimens brought from around the world to France before being sold to Scotland. Sapir Dvir shows how cars, maps, and guidebooks circulated between France and North Africa while enabling the movement of people, whether colonizers or tourists. In Della Zazzera's and Thiroux's articles, the gaslights and toilets obviously did not move, but they functioned because of the circulation of water and gas, and they shaped how people circulated through Paris streets. Della Zazzera's illumination burned gas delivered through a network of pipes unlike their predecessors, the oil lamps (*reverbères*) with their individual reservoirs of oil.

17. Murad and Zylberman, *L'hygiène dans la République*.
18. Zdatny, *History of Hygiene in Modern France*.
19. Hesse and Sahlins, introduction.
20. For French examples, see Aubin, "Fading Star of the Paris Observatory"; and Broch et al., "Moving Objects."

Thiroux's toilets, objects of intense concern, illustrate how French hygienists obsessed over the circulation of air, water, and waste as fundamental facets of their discipline, guaranteed by the careful design of drainage and ventilation.

Because artifacts traveled across networks of various kinds, the theme of networks threads through this issue in multiple ways. First were the overseas shipping networks that tied together France's empire and reached beyond it, for example, in the global search for scientific specimens. Steering flows of people, commodities, resources, and information in and out of France, ships, waters, and ports were the sinews of globalization both before and after steam power. Second, on a smaller infrastructural scale, networks of pipes routed water, gas, and other resources through Paris. Third were the social networks of scientific collaboration and communication, the ties that bound the republics of letters and of science. Finally, there were the revolutionary networks, both transimperial and radical, with important roots in Paris, that connected Louise Michel with the global community of the far left.[21] These varied types of networks overlapped but also played distinct roles. Shipping networks served entrepreneurs, colonists, specimen collectors, and tourists (not to mention bureaucrats and imperialists). The social networks of science, whether amateur or expert, facilitated exchanges of people, texts, specimens, and ideas beyond France. These networks linked France with peer institutions in Edinburgh and Vienna, endeared the Swiss chalet to French engineers and hygienists, and solicited data and specimens from practitioners operating worldwide. Among radical intellectuals like Michel, networks helped sustain a revolutionary community of scholars who believed that another science was possible.

The multiple mobilities illuminated in this issue comprise moving people and moving things, largely outside the domain of state-directed mobility of colonial expeditions, trade agreements, troop movements, and diplomatic exchanges. The ramifying global networks of France's expanding empire were crosscut by transnational and transimperial networks that suggest what French historians can learn by looking beyond France and its empire.

Decentering the State

Statism used to be state-of-the-art in French history and in French history of science. The contrast between Britain's independent Royal Society (1660) and France's state Academy of Sciences (1666) is a familiar one.[22] While centralization remains an important dynamic across early modern and modern French

21. Anderson, *Under Three Flags*, 174–76.
22. Fox, *Savant and the State*; Smith, "Longest Run."

history, this issue indicates a relative decentering of the state in histories of science and technology. Although these articles still show state involvement, the state neither drove nor determined the changes in science and technology explored in these articles. Not surprisingly, then, the authors also pursue science and technology outside of scientific institutions. Even though France's national museum of natural history remains prominent in "Exporting French Nature," Tsal makes clear that this institution could not have functioned without the help of a diverse global network of collaborators, both nonstate and non-French actors. In addition to formal public-private partnerships (such as state contracts or *concessions*), these articles also document informal collaboration of state and nonstate actors. In Sapir Dvir's article, tourists and sellers of cars, tires, guidebooks, and vacations contributed to France's colonization of north Africa without any formal partnership with the state; rather, they were inspired by patriotism, profit, the civilizing mission, a sense of adventure, and combinations thereof. While science and technology are important objects of humanistic study because they serve as tools of state power, some of that power is soft power. A related consequence of decentering the state is revising our sense of which institutions matter in histories of science and technology—both those outside the public sector (the state institutionally speaking) and those outside the nation and its empire (the state territorially speaking). Decentering the state works against narrow institutionalism and encourages transnational perspectives; both approaches are featured in this issue.

Another consequence of decentering the state is deemphasizing revolution and regime change. Although four of our six articles are nineteenth-century studies, the recurring revolutions of that century do not appear prominently in these articles. The violence of the Revolution, the Napoleonic years, and the Commune is always in the historical background, because significant moments in scientific and technological change do not line up with important moments of regime change. For example, what John Tresch called "Romantic" science and technology continued apace before and after the revolution of 1830, pre-Pasteurian hygiene extended before and after the cholera epidemics of 1832 and 1849, and Latour's Pasteurian hygiene continued across the breach of 1870.[23] The decentering of the state therefore indicates how finely grained histories of science and technology can revise our sense of chronology and periodization. This pattern is quite striking in the nineteenth century in which ostensible moments of historical rupture have tended to overdetermine how we divide the past and

---

23. Tresch, *Romantic Machine*; Latour, *Pasteurization of France*. See also LaBerge, *Mission and Method*.

whether we see continuity or discontinuity in historical narratives. Not all breaks in history correspond with crisis moments in state power.

Conclusion

These articles document conflicts and disagreements that haunted science and technology in France, in its empire, and beyond from the mid-eighteenth to the mid-twentieth centuries. Inevitably, they resonate with our ongoing culture wars, heirs to the "science wars" of the 1990s. Here we are not encouraging facile analogies, much less proposing naive lessons of the past. That said, if we agree that the study of history opens a space for reflection on concerns agitating us now, and if we cautiously pursue historical analogies, these articles provide fascinating fodder for thinking about current-day themes.

First, they urge us to think about how scientific and technological innovations, coming hard and fast, get wrapped up in and are often explicitly marketed for ideological ends. Technological hype characterized 1820s gaslight boosters, 1880s sewerage reformers, and 1920s automobile promoters. Yet why don't more of us ask, as the 1820s Romantics did, whether an innovation is merely new, rather than necessary—and perhaps even inimical to our society's well-being? These stories reveal familiar rhetorics of technological optimism and optimization often applied to artificial intelligence (AI) today. What does it say about us that our contemporary debate about AI is dominated by corporations that will profit the most from it and whose views are amplified by their control of digital communication and the social prestige accorded capitalist leadership?

Fraught questions of water and public hygiene link the Third Republic's public toilets for ladies with last summer's debate over cleaning up the Seine for Olympic swimming. Both sets of questions were inspired by the environmental impact of sewers from Paris and other cities upstream. In the summer of 2024, as Paris mayor Anne Hidalgo plunged into the river and emerged unscathed, cheeky Parisians were conspiring to defecate in it while news stories flickered that swimmers from Belgium, Germany, Ireland, Norway, Portugal, Sweden, and Switzerland got stomach bugs after swimming there. Mobilities abounded. As bacteria, athletes, and the mayor circulated in the river, rumors circulated in the media, animating contests over water quality, swimmer safety, local politics, and international relations. Amid all this conspicuous circulation, and as the world watched Paris, who was to decide which waters were clean or safe enough?[24]

The fact that the nation of Pasteur, once so proud of its role in developing modern vaccines, confronts a growing antivaccine movement in the age of COVID-19 urges us to historicize the faith, hope, and trust invested in science

---

24. For historical context, see Weiss, "Making Engineering Visible"; and Barnes, *Great Stink of Paris*.

and technology and the social power of expertise. As noted above, these articles frequently show individuals outside the charmed circle of professional experts who challenged them and even made science on their own. We now live in a world of citizen science movements, for such causes as environmental justice or health care reform, alongside antiscience movements for undermining vaccines or action against climate change. How do we balance skepticism, a democratic openness to views beyond the professionally sanctioned, and the intellectual authority of professionals, on whom we depend for the next vaccine and energy-saving innovations? More important, how do we encourage our students to do so?

Carlyle's article demonstrates particularly well how pedagogical programs intended to communicate scientific knowledge through new technologies in new venues generate conflicts as well as hopes, which inevitably connect with debates over values that, in their most acute form, spark moral panics. Finding contemporary examples is like shooting fish in a barrel, but these examples are complicated by the increasing flexibility and complexity of both bodies and behaviors, both sex and gender. In other words, we can expect conflicts over the nature of the body and the roles it allegedly suits us to become, if anything, more ubiquitous and intractable, pitting ideals of individual happiness or health on the one hand against fears of moral collapse on the other.

Our toggling between past and present here is no more exhaustive than our pursuit of this issue's themes, which resemble so many threads in a richly colored and textured fabric. Like the articles in this issue, it potentially suggests additional research questions about science, technology, modernity, and mobility in French and Francophone history. What other artifacts produced through scientific and technological innovation can scholars similarly leverage to excavate the moral stakes people in the past perceived in them? Innovators more often than not have swathed their innovations in promises of improvement, so it is the business of historians to ask "What improvement? Defined by whom, for whom, and to what end?" We have mentioned the decentering of the state and the role of marginality, allegedly justified on the basis of gender, race, and civilizational backwardness, in knowledge production. If Sapir Dvir's article decenters the state in French pursuit of imperial goals, what does it mean when transnational capitalistic entities begin to matter more in technological and scientific research? What is the impact on the promise of a more democratic production of knowledge? Can we be sanguine about such developments, as science and technology are arguably important drivers of social change, potentially affecting the lives of everyone on the planet? As these articles show, science and technology have never spoken for themselves in any historical moment. Rather, human beings continuously contested their meanings, assessed them in the contexts of their politics, their societies, their economies, their cultures, from the perspectives of their social situations, aspirations, and fears. All of this makes topics in these areas not just fruitful for French and Francophone historians but crucially important.

APRIL G. SHELFORD is associate professor emerita in the Department of History at American University in Washington, DC. Her book *A Caribbean Enlightenment: Intellectual Life in the British and French Colonial Worlds, 1750–1792* appeared in 2023.

PETER S. SOPPELSA is assistant professor in the Department of the History of Science, Technology, and Medicine at the University of Oklahoma. He is coeditor with Suzanne Moon of *History of Technology: Critical Readings*, 4 vols. (2020).

Acknowledgments

We offer our warmest thanks to Carol Harrison, the editors and staff of *FHS*, the referees and contributors to this issue for helping us make the project successful.

References

Anderson, Benedict. *Under Three Flags: Anarchism and the Anti-Colonial Imagination*. Verso, 2005.
Aubin, David. "The Fading Star of the Paris Observatory in the Nineteenth Century: Astronomers' Urban Culture of Circulation and Observation." *Osiris*, 2nd ser., 18 (2003): 79–100.
Barnes, David S. *The Great Stink of Paris and the Nineteenth-Century Struggle Against Filth and Germs*. Johns Hopkins University Press, 2006.
Bertucci, Paula. *Artisanal Enlightenment: Science and the Mechanical Arts in Old Regime France*. Yale University Press, 2017.
Bleichmar, Daniela, and Peter Mancall, eds. *Collecting Across Cultures: Material Exchanges in the Early Modern World*. University of Pennsylvania Press, 2011.
Broch, Ludivine, William G. Pooley, and Andrew W. M. Smith. "Moving Objects: French History and the Study of Material Culture." *French History* 37, no. 4 (2003): 345–56.
Chartier, Roger. "Sciences et savoirs." *Annales: Histoire, sciences sociales* 71, no. 2 (2016): 451–64.
Chevallier, Fabienne. *Le Paris moderne: Histoire des politiques d'hygiène, 1855–1898*. Presses universitaires de Rennes, 2010.
Clayson, Hollis. *Illuminated Paris: Essays on Art and Lighting in the Belle Epoque*. University of Chicago Press, 2019.
Creswell, Tim. *On the Move: Mobility in the Modern Western World*. London, 2006.
Delbourgo, James. *Collecting the World: Hans Sloane and the Origins of the British Museum*. The Belknap Press of Harvard University Press, 2017.
Doig, Kathleen Hardesty, and Felicia Burger Sturzer, eds. *Women, Gender, and Disease in Eighteenth-Century England and France*. Cambridge Scholars Publishing, 2014.
Dupré, Sven, and Geert Somsen, eds. "Forum: What Is the History of Knowledge?" *Journal for the History of Knowledge* 1, no. 1 (2020): 1–2.
Findlen, Paula, ed. *Empires of Knowledge: Scientific Networks in the Early Modern World*. Routledge, 2018.

Findlen, Paula. *Possessing Nature: Museums, Collecting, and Scientific Culture in Early Modern Italy.* University of California Press, 1994.

Freudenthal, Gad. "The Historian of Science's Guide to France." *Isis* 78, no. 4 (1987): 575.

Freudenthal, Gad. "Science Studies in France: A Sociological View." *Social Studies of Science* 20, no. 2 (1990): 353–69.

Fox, Robert. "Fashioning the Discipline: History of Science in the European Intellectual Tradition." *Minerva* 44, no. 4 (2006): 410–32.

Fox, Robert. *The Savant and the State: Science and Cultural Politics in Nineteenth-Century France.* Johns Hopkins University Press, 2012.

Gelbart, Nina Rattner. *Minerva's French Sisters: Women of Science in Enlightenment France.* Yale University Press, 2021.

Harvey, David. *The Condition of Postmodernity: An Enquiry into the Origins of Cultural Change.* Wiley-Blackwell, 1991.

Hesse, Carla, and Peter Sahlins. Introduction to "Mobility in French History," edited by Carla Hesse and Peter Sahlins. Special issue, *French Historical Studies* 29, no. 3 (2006): 347–57.

Jennings, Eric. *Imperial Heights: Dalat and the Making and Undoing of French Indochina.* University of California Press, 2011.

Jordanova, L. J. *Sexual Visions: Images of Gender in Science and Medicine Between the Eighteenth and Twentieth Centuries.* University of Wisconsin Press, 1993.

Kim, Mi-Gyung. *The Imagined Empire: Balloon Enlightenments in Revolutionary Europe.* University of Pittsburgh Press, 2016.

LaBerge, Ann. *Mission and Method: The Early-Nineteenth-Century French Public Health Movement.* Cambridge University Press, 1992.

Lachapelle, Sofie. *Conjuring Science: A History of Scientific Entertainment and Stage Magic in Modern France.* Springer, 2015.

Lachapelle, Sofie. *Investigating the Supernatural: From Spiritism and Occultism to Psychical Research and Metapsychics in France, 1853–1931.* Johns Hopkins University Press, 2011.

Laqueur, Thomas. *Making Sex: Body and Gender from the Greeks to Freud.* Harvard University Press, 1992.

Latour, Bruno. *The Pasteurization of France.* Translated by Alan Sheridan and John Law. Harvard University Press, 1998.

Latour, Bruno, and Geof Bowker. "A Booming Discipline Short of Discipline: Social Studies of Science in France." *Social Studies of Science* 17, no. 4 (1987): 718–45.

Levin, Miriam R. "What the French Have to Say About the History of Technology." *Technology and Culture* 37, no. 1 (1996): 158–68.

Levitt, Theresa. "Science and Technology Beyond the Barricades." *Technology and Culture* 54, no. 2 (2013): 377–81.

Lynn, Michael R. *Popular Science and Public Opinion in Eighteenth-Century France.* Manchester University Press, 2006.

MacClellan, James, III, and François Regourd. *The Colonial Machine: French Science and Overseas Expansion in the Old Regime.* Brepols, 2012.

Misa, Thomas, Andrew Feenberg, and Philip Brey, eds. *Modernity and Technology.* MIT Press, 2003.

Murad, Lion, and Patrick Zylberman. *L'hygiène dans la République: La santé publique en France, ou l'utopie contrariée (1870–1918).* Fayard, 1996.

Pestre, Dominique. "Pour une histoire sociale et culturelle des sciences: Nouvelles définitions, nouveaux objets, nouvelles pratiques." *Annales: Histoire, sciences sociales* 50, no. 3 (1995): 487–522.

Picard, Jean-François. "L'organisation de la science en France depuis 1870: Un tour des recherches actuelles." *French Historical Studies* 17, no. 1 (1991): 249–68.

Quinlan, Sean M. *The Great Nation in Decline: Sex, Modernity, and Health Crises in Revolutionary France, c. 1750–1850.* Aldershot, 2007.

Rieger, Bernhard. *Technology and the Culture of Modernity in Britain and Germany, 1890–1945.* Cambridge University Press, 2005.

Ross, Andrew Israel. "Dirty Desire: The Uses and Misuses of Public Urinals in Nineteenth-Century Paris." *Berkeley Journal of Sociology* 53 (2009): 62–88.

Schivelbusch, Wolfgang. *The Railway Journey: The Industrialisation of Time and Space in the Nineteenth Century.* University of California Press, 1977.

Schnapper, Antoine. *Collections et collectionneurs dans la France du XVIIe siècle.* Flammarion, 1988.

Schiebinger, Londa. *The Mind Has No Sex? Women in the Origins of Modern Science.* Harvard university Press, 1989.

Simon, Jonathan. "Forum Retrospectives: History of Science in France." *British Journal for the History of Science* 52, no. 4 (2019): 689–95.

Smith, Cecil O. "The Longest Run: Public Engineers and Planning in France." *American Historical Review* 95, no. 3 (1990): 657–92.

Sutton, Geoffrey. *Science for a Polite Society: Gender, Culture, and the Demonstration of Enlightenment.* Routledge, 2018.

Tresch, John. *The Romantic Machine: Utopian Science and Technology After Napoleon.* University of Chicago Press, 2012.

Vila, Anne C. *Enlightenment and Pathology: Sensibility in the Literature and Medicine of Eighteenth-Century France.* Johns Hopkins University Press, 1998.

Weiss, Sean. "Making Engineering Visible: Photography and the Politics of Drinking Water in Modern Paris." *Technology and Culture* 61, no. 3 (2020): 739–71.

Wilson, Lindsay B. *Women and Medicine in the French Enlightenment: The Debate over "Maladies des Femmes."* Johns Hopkins University Press, 1993.

Zdatny, Steven. *A History of Hygiene in Modern France: The Threshold of Disgust.* Bloomsbury, 2024.

Women and Anatomy Education in Enlightenment France
A Double-Edged Scalpel

MARGARET CARLYLE

ABSTRACT   This article examines the gap between the aspirational goals of eighteenth-century anatomy as a discipline meant to enfranchise women and the complex realities of their engagement with the subject, which elicited both admiration and fear. New materials in medical education, such as wax models and real skeletons, were perceived as both liberating tools and sources of moral anxiety. Women's interactions with these models and the anatomical knowledge they represented were often accepted and even celebrated when presented in an amateur context and under expert supervision. However, they were seen as troubling when consumed independently or in subversive settings. This tension culminated in a late-century backlash against mixed-gender visits to anatomical waxwork museums.

KEYWORDS   gender, medicine, waxworks, the body, morality

Anatomical waxwork museums were all the rage in eighteenth-century France and in the cultural cities of interest that French grand tourists visited. Florence was home to the most-feted such museum of the age, the Royal Museum of Physics and Natural History, better known as La Specola. Its display of life-size anatomies modeled in colored wax attracted a diverse cast of French travelers, from the astronomer Joseph Jérôme Lefrançois de Lalande (1732–1807) to the military surgeon René-Nicolas Dufriche Desgenettes (1762–1837).[1] The magistrate Jean-Baptiste Mercier Dupaty (1746–88) was completely spellbound by the waxworks and urged metaphysically minded philosophers to seek them out in order to understand "man" in all his complexity.[2] Women were also prominent visitors to La Specola, including the portrait painter Elisabeth Vigée-Lebrun (1755–1842).[3]

---

1. Lalande, *Voyage en Italie*, 568; Desgenettes, *Souvenirs*, 400–401.
2. Dupaty, *Lettres sur l'Italie*, 162–63.
3. Museum registers kept from 1783 to 1788 indicate that between a quarter and a half of La Specola's visitors were women. See Messbarger, "Re-Birth of Venus," 203.

French Historical Studies • Vol. 48, No. 2 (May 2025) • DOI 10.1215/00161071-11626753
Copyright 2025 by Society for French Historical Studies

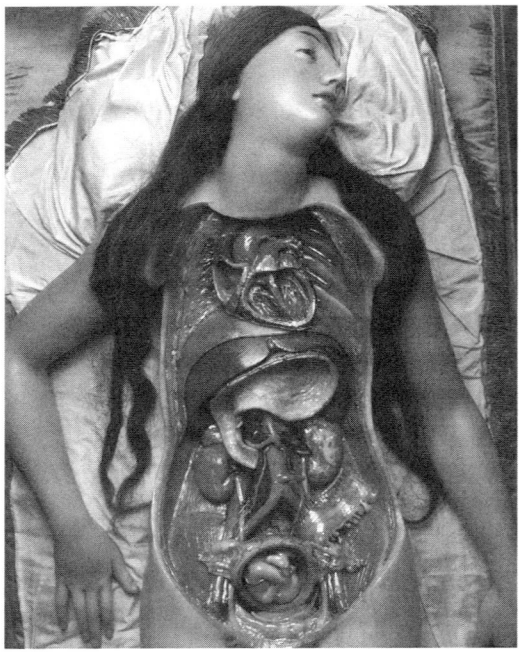

FIGURE 1 A view onto the interior organs, including fetus, of the life-size anatomical wax Venus at La Specola. Photograph credit: Saulo Bambi, Museum System of the University of Florence. Courtesy of the Museum of Natural History, La Specola, Italy.

A big part of the educational value of these museums lay in their ability to bring women and otherwise genteel audiences into the fold of human anatomy without offending their polite sensibilities. Waxworks were absent of the blood and stench of hands-on cadaveric dissection and yet lifelike enough to provide accurate anatomy lessons. Organs could be dissected out, allowing users to engage in a simulative-but-sanitized version of the manual work normally reserved for the male anatomist. This was achieved without any need to raise the scalpel and cut into rotting flesh. The female wax nudes known as "anatomical Venuses" depicting gravid wombs were widely recognized as the most cutting-edge examples of these medical models.[4]

The celebrated Venus was also the showstopper at La Specola, whose visitors interpreted it in a variety of registers. This particular Venus's supine pose, satin sheets, shock of pubic hair, and jewel-sized fetus invited medical, aesthetic, and sexual gazes, among others (fig. 1). Her finely crafted organs and elegant mahogany-and-glass display case placed her a register above the late-century fairground knockoffs or the models displayed in the *salons de cire* found in the entertainment districts of cosmopolitan cities like Paris.[5] The anatomical accuracy and technical virtuosity of the Venus could be lost on undiscerning eyes, however, and she was sometimes confused with her more pedestrian counterparts.[6]

4. Ceglia, "Importance of Being Florentine"; Ceglia, "Rotten Corpses," 435–41; Didi-Huberman, *Ouvrir Vénus*; Ebenstein, *Anatomical Venus*; Messbarger, "Re-Birth of Venus."

5. On fairground Venuses, see Hoffmann, "Sleeping Beauties"; and Wagner, "Replicating Venus," 13, 20–23.

6. It is important to remember that popular forms of entertainment were also styled as "improving." See Stephens, *Anatomy as Spectacle.*

Scholars of eighteenth-century anatomical models have acknowledged the range of responses they elicited from audiences and the sociocultural variables that conditioned their interpretation.[7] Anatomical models could be understood as titillating objects, but they also served as tools of inquiry designed to stimulate more sober imaginations. They occupied a position on the broad spectrum of eighteenth-century scientific learning, which spanned from popular entertainment to serious academic study.[8] Anna Maerker has observed that the classification of anatomical models as "medical technologies" was frequently contested due to their resemblance to those found in popular forms of entertainment. The artisan-anatomists who fashioned and marketed medical models for an up-market clientele thus worked hard to reaffirm their pedagogical value.[9] Elizabeth Stephens reminds us that efforts to make anatomical models like the Venus culturally legible to medical and lay audiences alike was a work in progress. The anatomical lessons embodied in waxworks were rarely self-evident, and audiences required instruction to "read" them in suitably scientific ways.[10] Scholars have also pointed to the disquieting aspects of wax anatomies: their uncanny lifelikeness was often noted by museumgoers, even in an age when death was more materially familiar than it is today.[11] These models could never quite shake their associations with death or sex, but that did not stop enlightened authorities from appropriating them as useful tools for educating a mixed-sex clientele in the science of the body.

My aim in this article is to analyze the moral tensions surrounding the use of these models in providing elite French women and girls with an education in human anatomy. I am most interested in the disconnect between the aspirational nature of eighteenth-century anatomists and educators to enfranchise women and the complex, often fraught reality of their encounters with the subject through object-based learning,[12] which elicited both adulation and moral fear. I have found that women's interactions with anatomical models were

7. A vast literature is devoted to eighteenth-century anatomical waxworks and wax modelers. See, e.g., Dacome, *Malleable Anatomies*; Hilloowala and Renahan, "XVIII Century Anatomical Models"; Lemire, *Artistes et mortels*; Messbarger, *Lady Anatomist*; Panzanelli, *Ephemeral Bodies*; Pirson, *Corps à corps*; and Schnalke, *Diseases in Wax*.

8. Lynn, *Popular Science*.

9. Maerker, "Anatomizing the Trade," 536–37.

10. Stephens, "Venus in the Archive," esp. 135.

11. Ceglia, "Rotten Corpses," 425–32; Simon, "Theater of Anatomy," 64. Much has been written about these models for their artistic and scientific value, including their craftsmanship and the properties of wax as a building material. See Ballestriero, "Anatomical Models and Wax Venuses"; Dacome, "Women, Wax, and Anatomy"; and Panzanelli, *Ephemeral Bodies*, 5.

12. On the role of material objects in learning and forms of "literacy," see Lehmann, "Material Literacy." See also Daston, *Things That Talk*. On anatomical objects in early modern science and art, see Carlyle and Reinhart, "Anatomical Things."

accepted and even celebrated when they were digested in amateur form and under expert guidance. But they were also seen as troubling and potentially subversive when consumed by the fair sex in unsupervised contexts. Models like waxworks served as a kind of double-edged scalpel in educational regimes, entreating girls and women to learn about their own bodies, whose depravities they were also not meant to know about. In sum, if these objects were seen as emancipatory teaching tools benefiting the "moral sex,"[13] their content also generated new forms of moral anxiety.

In what follows, I build a picture of the intrinsic moral slipperiness of medical models by examining the roles played by educators, philosophes, libertines, and anatomists in either promoting or proscribing their use as teaching tools. The first section explores how progressive pedagogues like Mme. de Genlis (1746–1830) furnished students with wider horizons that most notably included hands-on, object-based learning. She enlisted no-nonsense strategies to provide anatomical knowledge normally withheld from young men and women in a bid to emancipate them as they transitioned into adulthood. In this line of thinking, learning human anatomy on objects like real skeletons was a salutary pursuit that enhanced coming-of-age women's ability to navigate their future social roles as wife and mother. The designs of reformers like Genlis nonetheless raised alarm bells for conservative authorities, who feared amoral use of anatomical models when placed in the wrong hands. In the second section, I analyze the forms of moral panic that emerged around waxworks and how libertine authors like the marquis de Sade (1740–1814) deliberately played on them; he capitalized on contemporary perceptions of young women's innate impressionability to carry out sadomasochistic fantasies on waxworks.

In the third section, I tease out the tensions between these competing discourses by spotlighting women's ingenuity in creating, exhibiting, and consuming waxworks. If male luminaries like Denis Diderot (1713–84) encouraged women to become learners and realize their full potential as enlightened citizens first and as bodies with procreative roles second, it was his neighbor, the Parisian anatomical wax modeler Marie-Marguerite Biheron (1719–95), who built this knowledge from the ground up. Biheron pioneered lessons in reproductive anatomy using models of her own invention at a time when elite girls and women found expanding opportunities to educate themselves in human anatomy through courses and hands-on dissections. Finally, I take Elisabeth Vigée-Lebrun's firsthand account of her visceral reaction to a disemboweled wax Venus as an opportunity to assess the complex ways women's engagement with waxworks were "read" by Enlightenment culture. I use this example to foreground the fourth and final section.

13. Steinbrügge, *Moral Sex.*

where I scrutinize late-century debates over the benefits and dangers of women's entry into wax museums.

Skeletons in the Hallway: A Coming-of-Age Education

The anatomical waxworks displayed in public museums provided a pathway for women to gain an education outside the traditional spaces of French scientific learning typically reserved for men, such as the Paris Faculté de Médecine, the Académie Royale des Sciences, and the Académie Royale de Chirurgie.[14] In these hallowed venues, the subject of human anatomy was the prerogative and business of learned men. Faculty professors trained future cohorts of male surgeons and physicians in human anatomy, while leading academicians circulated descriptions of unusual cases and experimental reports on human bodies. Male savants broached a range of topics, from the nature of hermaphroditism[15] to how an enslaved Black woman might conceive twins of different races.[16]

The tides of education were nonetheless turning. Women were increasingly recognized as invaluable students and consumers of science, a process that began in the late seventeenth century, when the first textbooks aimed at educating elite women in such subjects as astronomy and botany were published.[17] This literary tradition was rounded out in the mid-eighteenth century with the proliferation across Paris and the provinces of numerous free and for-fee public courses in science, including chemistry, natural history, and physics.[18] Women also actively partook in natural philosophy, as science was known then, as collectors and connoisseurs of curious objects,[19] as well as truth-seeking tourists who entered museum spaces. Women's opportunities to pick up scientific subjects, including anatomy, thus expanded in the name of public science aimed at mixed-sex elite audiences.[20]

Did the process of bringing female audiences into the fold of anatomy learning apply to school-age girls? Human anatomy was not a subject traditionally conveyed to young women in early modern France, though that was true of

14. On the gendering of scientific spaces, see Terrall, "Gendered Spaces."
15. See, e.g., Bibliothèque de l'Académie Nationale de Médecine, Académie Royale de Chirurgie, carton 9B, dossier 25, "Espèce d'un hermaphrodite," signed July 10, 1781, by Loche; carton 9B, dossier 141, "Observation sur une prétendue hermaphrodite" signed 1778 Lauverjat and report signed 1779 Peltier; carton 9B, dossier 77 "Ouverture d'un hermaphrodite," signed Bernard.
16. Bibliothèque de l'Académie Nationale de Médecine, Académie Royale de Chirurgie, carton 25, dossier 77, no. 163, "Idées générales sur la génération," signed [n.n.] 1774.
17. Douglas, "Popular Science"; Terrall, "Natural Philosophy."
18. For a list of anatomy courses in Paris, see *Etat de la médecine*. See also Huard, "L'enseignement medico-chirurgical"; and Gelfand, "'Paris Manner' of Dissection," esp. 103–11.
19. Carlyle, "Collecting the World in Her Boudoir."
20. Sutton, *Science for a Polite Society*; Walters, "Conversation Pieces."

most scientific subjects.[21] This began to change when enlightened pedagogues like Genlis came onto the scene. In 1782 she was appointed the first female educator or *gouverneur* of princes. No sooner had she and her young royal charges moved to the Bellechasse royal residence on the Rue Saint-Dominique in Paris than she set to work on reimagining their education.[22] Her pedagogical outlook was fairly traditional, accepting sex and social standing as key factors in determining curricula, but she did embrace the forward-thinking philosophies of Etienne Condillac (1715–80), who privileged sensory experience in education.[23] In the same year she became royal governess, she published *Adèle et Théodore ou lettres sur l'éducation*, a fictional treatise in epistolary form that served as a platform to elucidate her pedagogical commitments.[24] The character of the baronne d'Almane develops a curriculum for the two eponymous youngsters of each sex, which she carries out with the assistance of tutors. This mirrored the curriculum that Genlis set for her own students, which prioritized practical everyday experiences and forms of manual apprenticeship over bookish rote learning. At Bellechasse, her royal charges cultivated botanical gardens while conversing with a German tutor and were accompanied by an apothecary on country walks to study medicinals.[25] The royal children absorbed the décor around them, including the tableaux, tapestries, and maps of the château. They were taught science using magic lanterns and games and were taken to workshops to learn arts like joinery firsthand.[26]

Adèle et Théodore provided like-minded tutors of young nobles with tips on how to incorporate sensationalist philosophy into their lesson plans, even, and especially, for difficult subjects that naturally invoked fear in students. The baronne d'Almane's self-proclaimed goal as an educator was "to familiarize my children with all the things that can naturally inspire disgust and fear." She asserted that young girls and boys should be trained to overcome these fears by confronting them head on: embolden them to handle creatures of the living world—frogs, spiders, mice—and actively partake in nature's frightful spectacles, such as thunder and lightning.[27] Through the character of the baronne,

21. On female education in Enlightenment France, see esp. Sonnet, *L'éducation des filles*. On early Enlightenment education, see Gill, *Educational Philosophy*, 65–116.

22. On her life, see Genlis, *Adelaide and Theodore*, ix–xx; and Naudin, "Stéphanie-Félicité, Comtesse de Genlis."

23. See, e.g., Riskin, *Science in the Age of Sensibility*, esp. 1–18. On the education of princes, see Chartier et al., *L'éducation en France*, 178, 181, 199–200; and Halévi, *Le savoir du prince*, 53–56.

24. Genlis, *Adèle et Théodore*.

25. Véron, *Mémoires d'un bourgeois*, 57–59.

26. Genlis commissioned a series of scale models constructed in 1783 for teaching purposes. The resulting miniature ateliers, depicting woodworkers, chemists, and locksmiths, among others, were inspired by the engravings of the industrial arts in the *Encyclopédie*. See Burger, *Les maquettes*.

27. Genlis, *Adèle et Théodore*, 142. All translations from French into English are my own.

Genlis advocated exploring the range of sense perception, with sight emerging as particularly valuable in providing a distinct and clear impression to students.[28] Visual and material culture also allowed the baronne to sensitize her two young-sters to human anatomy through a simple yet effective exercise involving a real skeleton.

Genlis advised tutors to cultivate in their charges an indifference to the taboo nature of human anatomy through forms of demystification aimed at the domestication of such knowledge. In an exercise designed to make the human body commonplace and familiar, rather than forbidden and terrifying, a skele-ton was the key teaching prop. The baronne placed "a large glassed-in armoire through which one can see a skeleton and several anatomy specimens" in a cor-ridor through which Adèle and Théodore walked regularly. Lest the presence of a skeleton in the hallway incite "a first nasty impression" that is "ever difficult to overcome," the governess recommended primer exercises designed to ease the children into the subject. Among the measures she took was a dinner table con-versation about "the various anatomy pieces that had been sent to me from Paris." A convenient dinner guest, an anatomist named Aimeri, was on hand to provide particulars. His own passion for the subject was so profound that "for two years, his bedroom had been entirely filled with skeletons."[29]

The children's introduction to anatomy at the dinner table had the desired effect: they began asking questions, even sometimes the seemingly wrong ones, but always with a view to arriving at complete and sound knowledge. Adèle pre-emptively concluded "that a skeleton must be a very villainous thing" only to change her mind once the dinner party retired to the hallway, where she and her brother "witnessed the skeleton with neither surprise nor disgust." The baronne's lesson worked. From this moment on, the children "passed through this corridor all the time without imagining that one could harbor any fear of a skeleton."[30] She had also transformed the skeleton, that time-honored symbol of the finality of death, into a useful teaching tool.[31]

Tactile objects like anatomical waxworks also allowed impressionable young minds to vanquish their innate horror. The brand of waxworks that emerged by the second half of the eighteenth century embodied healthy organs that could be removed and replaced by users.[32] They rendered the body's hidden and bewil-dering inner workings as a novel form that insisted on the vivacity of life, not

28. Genlis, *Adèle et Théodore*, 45.
29. Genlis, *Adèle et Théodore*, 143.
30. Genlis, *Adèle et Théodore*, 144.
31. On the history of the early modern skeleton, see Guerrini, "Whiteness of Bones."
32. Ceglia, "Rotten Corpses," 431–35. On the role of touch in interactive anatomical displays, see Maerker, "Towards a Comparative History of Touch."

the inevitability of death. Life-size whole bodies, as well as stand-alone organs and systems, were displayed in purpose-made display cabinets in museum spaces where their overseers ensured controlled, polite encounters with them. The wax-works on offer in such exhibitions thus allowed ladies and gentlemen interested in dabbling in anatomy with genteel alternatives to the real cadavers normally studied by male medical professionals.[33]

Lessons on anatomical waxworks were also described in educational litera-ture. In Mme. Jeanne-Marie Leprince de Beaumont's (1711–80) *Magasin des ado-lescentes* (1760), a scene unfolds in which a governess named Mme. Bonne pro-vides an overview of the mechanics of the digestive process to two young noble girls, Lady Louise and Lady Violente. The governess announces that the next step was to observe this process in the book of nature, but both ladies recoiled in fear over the prospect of taking in such a ghoulish spectacle as a human cada-ver. Mme. Bonne reassured them that they could substitute a dissection with artificial teaching aids: "You can spare yourselves some part of this repugnance. Showing in London are anatomical figures in wax, which are exactly like the human body."[34]

Mme. Bonne may have been referring to the real-life waxworks created by the acclaimed anatomist Guillaume Desnoües (1650–1735). His models were dis-patched to London from Paris in 1718 before going on tour across the British Isles and eventually finding a permanent home at Trinity College in Dublin in 1753. In Desnoües's absence, his models were regularly hired out to male surgeons and man-midwives, who used them as teaching tools.[35] One of Desnoües's most popular teaching aids was a wax replica of a full-term parturient who had died while delivering a fetus that was still trapped in her womb. He preserved the original maternal-fetal body using embalming and wax injection techniques and debuted it to an audience of some two thousand surgeons, students, and nobles in the Great Hall at the Hospital of Genoa.[36] Desnoües's models were clearly an inspiration, leading to several imitations in London. Among these imitators was the surgeon-anatomist Abraham Chovet (1704–90), who enhanced his life-size artificial model with a mechanism that pumped blood-colored fluids through the veins and arteries of both mother and fetus.[37]

---

33. Craske, "'Unwholesome' and 'Pornographic,'" 78; Dacome, "Women, Wax, and Anatomy," 526–27.

34. Beaumont, *Magasin des adolescentes*, 145.

35. Craske, "'Unwholesome' and 'Pornographic,'" 80. Newspaper advertisements in the 1730s indi-cate that the models were on loan from their owners and could be viewed for a fee. See, e.g., *London Post*, 1.

36. Desnoües and Guglielmini, *Lettres de G. Desnoues*, 31–33.

37. Register Book Original (hereafter RBO), Royal Society, Abraham Chovet, "Proposals for Making Anatomies of human Bodies in Colour'd Wax After the Manner of Those Belonging to the Sieur Desnoues, by Abraham Chovet, Surgeon (1732)," RBO/17/33; RBO, Royal Society, Abraham Chovet, "A Proposal to Make Anatomies of Human Bodies in Coloured Wax by A. Chovet," 1732, JBC/14. My thanks to Keith Moore for providing these records.

Sex Education: The Dangers of Dirty Books and Boudoirs

Desnoües's birthing-scene waxwork captivated and startled audiences. It also shed light on a troubling side of anatomical models: the truth of sex. Even as they became a mainstay of elite scientific sociability and enlightened education, thanks to the object-based teaching methods championed by reformers like Genlis, medical models remained contentious due to the access they provided to sexual knowledge. The first generation of anatomical wax models, including those fashioned by Desnoües, were neither inherently virtuous nor immediately recognized for their educational utility. In this section, I show how incorporating models of reproductive anatomy into education schemes aimed at young women was hotly contested. These simulacra, which promised to provide a rational approach to anatomical education, were met with skepticism by conservative thinkers, who viewed them as a cover for immorality and an easy outlet for debauchery. Their apprehensions were fueled in no small part by the proliferation of pornographic literature featuring fictionalized female libertine "learners" who trivialized anatomy lessons in their boudoirs.

Anatomical models depicting sex acts or their literal products, fetuses, faced particular scrutiny, requiring concerted efforts to detach them from any sense of moral ambiguity and integrate them into the mainstream of polite learning. Even with such efforts, widespread acceptance for education in reproductive anatomy remained elusive. Desnoües, who pioneered medical moulage with his collaborator, Gaetano Giulio Zumbo (1656–1701), discovered this firsthand when royal approval for his novel brand of waxworks failed to protect him from censure. In 1712 the morality of Desnoües publicly displaying his models became a focal point of debate. This overshadowed the significant scientific value of his fee-earning wax museum, which opened for business in January 1711 after he received a royal privilege that was converted into letters patent that July. These deeds granted him the right to exhibit his waxes before the public "in the city of Paris and elsewhere and to use them for anatomical demonstrations."[38]

It quickly became apparent, however, that the curious who indulged in his displays risked being morally compromised by what they saw. In response to complaints about the propriety of Desnoües's displays, the Parlement of Paris issued a court order in August 1712 imposing restrictions on the museum's opening hours. Desnoües was thereafter prohibited from conducting demonstrations on his models in the evenings. The order also stipulated that the "natural parts

---

38. Archives Nationales, O¹ 55, fol. 223, "Privilège pour faire des ouvrages anatomiques de cire coloriée Paris."

of both sexes will always be covered." Additionally, he was required to prominently display the court order on both sides of the entryway and above any interior space where demonstrations took place, so that "nobody can feign ignorance" of the new restrictions. Failure to comply would result in a sizable 1,000
livres tournois fine and the confiscation of his royal privilege and wax figures.[39]

To be sure, some of the objections to the anatomical knowledge unveiled
in Desnoües's early-century waxworks had been mitigated by the time that public museums like La Specola opened their doors during the last quarter of the
century. It helped that waxworks had by this time generated the attention of
Europe's elite, who commissioned models for grandiose educational purposes.
For instance, Austrian Emperor Joseph II (1741–90) paid handsomely for replicas of La Specola's models, which were transported across the Alps to Vienna
and displayed to the public as a form of citizen education.[40] The initial objections to the sexual content of anatomical wax models were nonetheless difficult
to overcome, even when the waxworks were visually and spatially dressed up as
teaching tools intended for enlightening purposes.

Questions that consistently arose were who ought to have access to these
models and for what purposes. Authorities saw sexual anatomy as the preserve
of male medical personnel, reasoning that it might corrupt young minds, particularly those of girls. Not all parents shared the progressive views of Genlis, who
believed that coming-of-age girls were just as suited as boys to anatomical study.
During the revolutionary period, a certain "Ledoux" satirized his daughter Judith's
obsession with anatomy. He noted with disapproval that she had decorated her
bedroom with "a damn skeleton" as well as "*écorchés*, skulls, and bones, each
more disgusting than the last." While he dismissed her devotion to anatomy as "just
another mania," he was concerned that her mother's premature death meant that
she would not learn how "to sew, to embroider, to mend my shirts." It did not help
that she has "a slightly weak brain, and from time to time a touch of madness" that
he saw as common to artists.[41]

If a young girl's unbridled collection of anatomical paraphernalia was
cause for concern, then her access to libertine literature surely presented a more
alarming flirtation with the subversive. Women were depicted in transgressive
erotic literature as constitutionally feeble creatures who were susceptible to
male seduction, moral corruption, and sexual vice.[42] The wild popularity and
increasing availability of such literature in the eighteenth century stoked fears

39. [n.a.], *Arrêt de la Cour de Parlement*. The *arrêt* is reproduced in full in Guiffrey, "Le cabinet anatomique," 163–64.
40. Maerker, *Model Experts*, esp. 19–49.
41. Ledoux, "Peinture," 1–3.
42. Harvey, *Reading Sex*, 102–23.

that sexual acts could be styled to naive audiences of young women as genuine knowledge-seeking exercises. It did not help that the same kinds of props that were used to enlighten adolescents about the human body were integral to the unsavory sexual acts described in such literature. If patrolling Desnoües's late-night waxworks viewings had been a challenge, it was even more difficult for censors to comprehensively police what women did to and with their own bodies.

Even some of the most outwardly enlightened figures, most notably Denis Diderot, who championed sexual education and authored such erotic works as *Les bijoux indiscrets* (1748) in which "talking vaginas" drive the storyline, feared the loss of parental and moral control when his own daughter was first exposed to libertinism.[43] One day, while walking through her bedroom, he caught her laughing uncontrollably and asked her what was so funny:

> –I laughed at Doctor Pangloss, who was giving lessons in experimental physics to Mme Paquette in a grove.
>> –What! You are reading *Candide*!
>> –Yes, dear dad, it is an infamous book, but since I have already started reading it, you will allow me to finish it.
>> –And who lent you this book?
>> –Ah! Dear dad, don't get involved, it's my business, and you can rest assured that this man will not have slighted me with impunity. . . .
>> –Another woman would have hidden the book, would have poisoned her imagination; my daughter did none of these things.[44]

The scandalous passage to which Diderot's daughter referred could be found in the opening pages of Voltaire's (1694–1778) satirical novella *Candide* (1759), in which Cunégonde, the malleable young female love interest of the eponymous male hero, witnesses a curious and memorable "lesson in applied physiology" while strolling near the castle in a small wood. In voyeuristic fashion, Cunégonde "caught a glimpse through the bushes of Dr. Pangloss" giving his lesson to her mother's maid Paquette, "a very pretty and very receptive little brunette." Cunégonde "noted in breathless silence the repeated experiments to which she was witness." She thus plays the role of the overly curious young woman who bears witness to a raunchy sex act that is veiled as an introduction to "applied physiology" and performed by none other than a syphilitic maid and her lecherous male tutor.[45]

This memorable opening scene serves as both a pretense for the young maid and her male tutor to explore their passion and an unintended lesson in sex

---

43. Diderot, *Les bijoux indiscrets*. On Diderot's view of the corporeal in Enlightenment discourse, see Goodden, *Diderot and the Body*.

44. Tourneux, *Diderot et Catherine II*, 392.

45. Voltaire, *Candide*, 4.

education for the impressionable Cunégonde.[46] The scene also satirizes the doggedness of science's increasing claims to objectivity: repeated "experiments" in the field produce nothing of utility, save babies, and might also cultivate a dangerous brand of female curiosity.[47] The consequences of indulging in this curiosity are universally deleterious: Cunégonde, once chastely naïve, is now perversely voyeuristic and irreparably corrupted. Meanwhile, the exuberant chambermaid-turned-pupil Paquette has singlehandedly spread Old World syphilis across the New World and, in so doing, has duped her supposedly all-knowing tutor, Pangloss.[48]

This scene in *Candide* reads like a page from a contemporary libertine novel littered with scenes of sexual apprenticeship that are integral to pornography as a genre.[49] Like Voltaire's *Candide*, erotic literature offered social critiques by insinuating that an understanding of sex comes from practice rather than any moral catechism or church confessional. The hands-on anatomy lessons that were valued in enlightened educational discourse for their sensory exploration were equally prized in the boudoir, but here the aim was sexual rather than intellectual enrichment. The trope of wicked, diseased male libertines corrupting young female minds was a staple in erotic literature from the mid-seventeenth to late eighteenth centuries. As Margaret C. Jacob has demonstrated, the well-known French erotic novel *L'école des filles, ou la philosophie des dames* (1655) was framed as a dialogue between two female cousins, one of whom is an expert anatomist who recounts a sexual encounter, highlighting the perceived perils of giving women unrestricted access to scientific knowledge.[50]

There were other similarities in how libertine and scientific literatures of the age sought to engage inquisitive young female minds. In adopting a dialogical format and by inscribing the pretense of "schooling" and "philosophical education" for young women in its title, *L'école des filles, ou la philosophie des dames* played with the conversational conceit that was central to Bernard le Bovier de Fontenelle's (1657–1757) contemporary *Entretiens sur la pluralité des mondes* (1686) and the genealogy of scientific textbooks aimed at enfranchising polite women into the natural sciences.[51] This approach aimed to make complex scientific

46. J. G. Turner discusses the sexual initiation of a youth by a mentor as a signal device of the erotic-didactic literary genre ("Sexual Awakening as Radical Enlightenment"). On the concept of sex education, see Agin, "Sex Education."

47. On female curiosity, see Benedict, *Curiosity*, 118–57; and Gargam, "Between Scientific Investigation and Vanity Fair."

48. Voltaire, *Candide*, 10. Pangloss explains to Candide that he contracted syphilis from Paquette, who had caught it from a Franciscan monk, down a direct line from Christopher Columbus.

49. Abramovici, "Comedy of Ignorance," 207.

50. Jacob, "Materialist World of Pornography," 166.

51. On women as readers of Fontenelle, see Douglas, "Popular Science," 1–4.

concepts more accessible and palatable to young women. The vogue for introducing the fair sex to "philosophy" was parodied in libertine texts, with Sade painting a macabre scene in *La philosophie dans le boudoir* (1795) in which one Mme. de Mistival, mother of the young female libertine Eugénie, is penetrated by a male gardener infected with syphilis. She is then daintily stitched up to secure her diseased fate, symbolically confirming that Eugénie has completed her adolescent apprenticeship in sadomasochism.[52]

The ultimate lesson imparted on any young female reader who dared turn the page, like Diderot's daughter, was that an education in sexual anatomy can be picked up as casually in a garden grove as from a book. In so doing, libertine literature endowed otherwise innocent and familiar spaces with a new power of perversion, a kind of sensationalist anatomy lesson in the garden gone wrong. It also suggested that once sexual acts were unleashed into the wilds of the innocent young imaginary, these children became irredeemable. Such violation could not be undone, and the particularly pliant curiosity of girls made the introduction of the topic of sex that much more dangerous. One might be tempted to interpret the much-discussed last line of *Candide*, "Il faut cultiver notre jardin," as a call to cultivate one's sexual garden, that one true act of human self-preservation. In this interpretation, the garden metaphor is oriented toward male reproduction: to spread one's seed or sow one's wild oats. However, if we read the final line in light of the opening scene, in which the lowly maid spreads syphilis across the New World, we can locate a commentary on women's inability to interact with the science of the body outside two limited, and often contradictory, roles: first, as fictional pupils in disease-ridden libertine scenes, and second, as tried-and-true vessels of procreation.

No wonder libertine literature alarmed conservative-minded authorities. It reimagined what happens when forms of moral discipline and social control failed to keep women's sexuality in check: their natural prurience and innate moral weakness easily lead them to vice. By featuring heroines whom readers might wish to imitate, libertine texts provided a troublingly amoral source of information about anatomy and the mechanics of sexual intercourse. Many staple erotic works in the eighteenth-century "pornographic flood" of literature feature female libertines in starring roles, including the Marquis d'Argens Jean-Baptiste de Boyer's (1704–71) *Thérèse philosophe* (1748) and *Les nonnes galantes* (1740).[53] Such works trespassed well beyond the confines of the moralistic and medical instruction provided in earlier works, such as Nicolas Venette's (1633–98) popular manual *Tableau de l'amour conjugal* (1st ed. 1696).[54]

52. St-Pierre, "Sade's System of Perversity," 48.
53. Muchembled, *Orgasm and the West*, 115–55; Darnton, *Forbidden Bestsellers*.
54. Porter, "Love, Sex, and Medicine."

Pornographic literature also appropriated familiar facets of polite society, such as the feminized boudoir, to transform them into spaces of sin in which the active female imagination was left to run wild.[55] The anatomy cabinet was also a space ripe for corruption. In a perverse scene in *Les cent-vingt journées de Sodome* (1785), Sade drew inspiration from his own visit to La Specola in 1775, which he described in his more mainstream travel memoirs as merely an educational experience.[56] Herein the 52nd of the 150 passions that he files under the most extreme "third class of criminals" unfolds in a cabinet of anatomical wax corpses. But this was no ordinary display: rather than giving off the healthy glow that was typical of lifelike wax models, Sade's artificial corpses were tortured into submission. The male protagonist first "receives this girl in his cabinet chockfull of cadavers in wax, very well imitated" before letting her in on the real reason why she is here: he asks her to select from among the brutalized waxworks the method of her own death.[57]

Women in the Wax Museum

This frightening episode highlights how waxworks could just as readily be used for lessons in sexual debauchery as for more thoroughly wholesome and edifying agendas. Was Sade's heroine quite simply the author of her own demise? Were she and her libertine-inspired female brethren beyond redemption? Left in the hands of a male writer like Sade, the answer to both questions is likely a resounding yes. However, for well-to-do women eager to study anatomy without the intermediary of a male tutor, thankfully, there was Marie-Marguerite Biheron. She authored dozens of lifelike waxworks that she displayed to fee-paying visitors in her Parisian anatomy "cabinet," a space within her residence.[58] Her obstetrical models in particular gained the acclaim of the Académie des Sciences, the Académie de Chirurgie, and the Faculté de Médecine.[59] Biheron also came highly recommended by society women like Genlis, who urged the young comtesse de Coigny (1753–75) to satisfy her remarkable "passion for anatomy" by enrolling in

55. Delon, *L'invention du boudoir*; Saisselin, "Space of Seduction."
56. Casamaggi and St-Martin, "Sade à Florence."
57. Sade, *Œuvres complètes*, 147–48.
58. There is a growing literature on Biheron's anatomy career. See, e.g., Boulinier, "Une femme anatomiste"; Carlyle, "Artisans, Patrons, and Enlightenment"; Dacome, "Intimate Connections"; Gargam, "Marie-Marguerite Biheron"; and Rattner Gelbart, *Minerva's French Sisters*, 166–207.
59. On Biheron's demonstrations of her waxworks in 1759, 1770, and 1771, see Académie Royale des Sciences: Pochette de Séance, "Extrait des Registres de l'acadᵉ R. des sciences du 23 juin 1759," dated June 23, 1759; Pochette de Séance, "[Report on Mlle Biheron's Artificial Anatomy]," signed Morand, Delassone, Tenon, dated Feb. 14, 1770; and Pochette de Séance, "Discours prononcé à L'academie royale des sciences le 6 mars 1771, en presence de Sa Majesté le Roi de Suede," by D'Alembert, dated Mar. 6, 1771. On her visits to the Académie de Chirurgie and Faculté de Médecine, see [Biheron], *Anatomie artificielle*.

Biheron's courses. Genlis explained that Biheron gave anatomy lessons in her home on "whole anatomical subjects" made of her proprietary blend of wax and rags. The comtesse might also venture into the back garden to observe Biheron dissecting cadavers in a glassed-in cabinet that the anatomist playfully referred to as "her little boudoir."[60]

Biheron offered private courses to customers like Genlis and Coigny, who sought more hands-on tuition in anatomy than what public courses available to women typically provided.[61] According to her advocate Diderot, Biheron instructed dozens of elite girls, even more society women, and several prominent men of letters throughout her lengthy career. His confrere, Jean-Baptiste le Rond d'Alembert (1717–83), confessed to having learned more "real anatomy" during an eight-day course under Biheron than in six months under the renowned anatomist Antoine Ferrein (1693–1769).[62] Biheron also provided specialized lessons in reproductive anatomy, including for Diderot's daughter in advance of her wedding night, and received a court commission from Catherine the Great (II, 1729–96).[63] By all accounts, her lifelike models were widely acclaimed as morally respectable ways for women to learn anatomy, earning praise from progressive educators and natural philosophers alike, including Philibert Riballier (1710–91) and Charlotte Catherine Cosson de La Cressonnière (1740–1813), who recommended Biheron's no-frills lessons to all young women.[64]

Safeguarding modesty and decorum proved to be significant hurdles in justifying female education in anatomy, particularly in reproductive anatomy, due to its associations with libertinism. Furthermore, efforts to present waxwork anatomies as suitable for female audiences did not guarantee uncomplicated reactions to them. Women's contemplation of waxworks in mixed-sex display spaces like Biheron's were hardly textbook encounters. If anything, forms of policing and self-censorship escalated in anatomy exhibitions as the eighteenth century wore on, partly in response to growing fears around coeducation. Recordings of women's encounters with wax models are rare and thus are of special interest in shedding light on the female experience. The portrait painter of Queen Marie-Antoinette (1755–93), Elisabeth Vigée-Lebrun, found wax models as troublesome as they were enlightening.

In the spring of 1792, while on tour in Florence, Vigée-Lebrun entered La Specola in the company of its director and court physician, Felice Fontana (1730–1805). She recalled in her memoirs entering the museum an innocent, not

60. Genlis, *Mémoires inédits*, 253–54.
61. See notices on her private courses in *Avant-coureur*, 768; and *Etat de la médecine*, 230.
62. Diderot, "Sur le même sujet," 613–14.
63. Carlyle, "Artisans, Patrons, and Enlightenment," esp. 38–39.
64. Riballier and La Cressonnière, *De l'éducation physique*, 323–25.

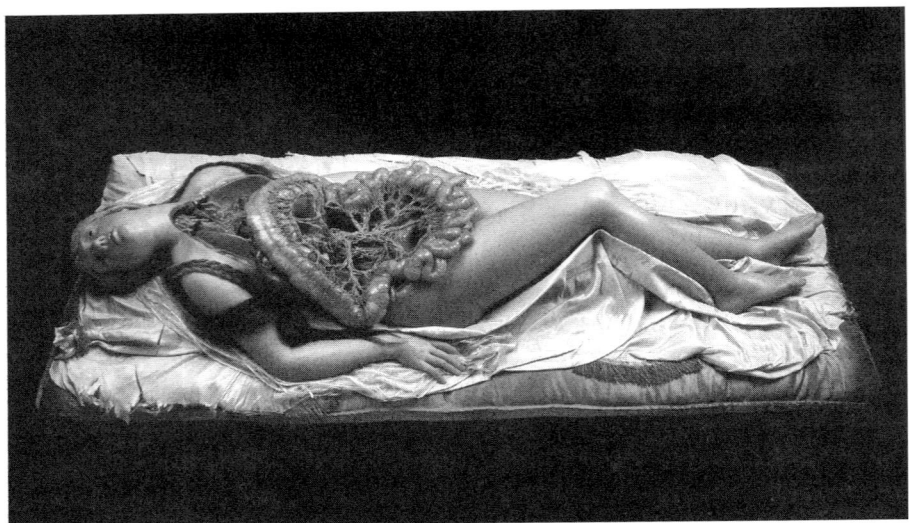

FIGURE 2 The disemboweled Venus examined by Vigée-Lebrun during her visit to the anatomical waxworks at La Specola. Photograph credit: Saulo Bambi, Museum System of the University of Florence, Courtesy of the Museum of Natural History, La Specola, Italy.

previously having seen anything "that would have made me experience a painful sensation," and leaving it disturbed after catching a glimpse of "a sleeping woman of life size" that Fontana beckoned her to approach. As she did, he lifted the exterior cover to reveal "all the intestines, turned as in ours," a sight that left her "in a deplorable nervous state" (fig. 2).[65]

It was the wax Venus's uncanny likeness to a real eviscerated body and her multisensory response to it, rather than any sense of its indecent presentation, that haunted Vigée-Lebrun most. The next time she saw Fontana, she sought his counsel on how "to divest [herself] from the inopportune susceptibility of [her] organs." She added that "I have heard too much. . . . I see too much and I feel all within a league."[66] At a time when individual human organs were still believed to be the seat of specific passions and diseases, Vigée-Lebrun's visceral reaction to the dissected Venus could be attributed to female hypersensitivity or constitutional weakness.[67] Her reaction was certainly in line with prescribed ladylike behavior, as much as Fontana performed appropriately in his gender role as a fearless male tutor. In the manner of a scientific showman, he heroically pulls

65. Vigée-Lebrun, *Souvenirs*, 238.
66. Vigée-Lebrun, *Souvenirs*, 238.
67. Anne C. Vila discusses how the medical tradition of correlating women's soft wombs and tender "cerebral pulp" with inferior mental capacities began with Pierre Roussel in the 1770s and continued with Pierre Cabanis into the Napoleonic period ("'Ambiguous Beings'").

back the Venus's modesty veil to reveal her inner workings.[68] In the absence of his own account, it is difficult to gauge Fontana's take on this episode, but there is no doubt that he intended to engage her in a memorable experience.

Vigée-Lebrun's encounter with the disemboweled Venus underscores that it was only natural that the cold realism of anatomical models evoked affective responses from viewers. A genteel woman being stimulated by the graphic nature of these objects was a salutary, indeed, a thoroughly human response. But affective reactions to waxworks may have gone too far when they struck a paralyzing fear in audiences, as they did for Vigée-Lebrun. The notion that excitement could lead to debility was an eighteenth-century medical trope, though it was not an ailment deemed exclusive to women.[69] Men of science wrote of being shocked by the stench emanating from cadavers on dissecting tables, while Sade was so appalled by the odorless plague-themed wax tableaux of Gaetano Zumbo that he brought his hand to his nose in a gesture of self-defense.[70] Ladies nevertheless remained the most likely candidates for succumbing to the inherent "terror" generated by "cold copies" of the human form. At least, that was according to the architect Athanase Détournelle (1766–1807), who advocated the closure of the Swiss-born modeler Philippe-Guillaume Mathé Curtius's (1737–94) popular Parisian waxworks, citing as the rationale the example of a woman who fainted of fright after encountering a lifelike statue while "promenading in the Chinese Garden."[71]

In this example, as much as in Vigée-Lebrun's account of her tour of La Specola, women exhibit the kinds of corporeal and mental frailty that were to be expected in the face of such displays. It would indeed be more remarkable if lifelike waxworks or statues did not speak to any woman, or man, of feeling. Nor could any degree of sensitization immunize women from visceral reactions to these models, reactions that were, in one sense, ungendered in an age that privileged affective responses to novel experiences. But there were limits. In contrast to the measured response of Genlis's imagined youngsters to the hallway skeleton, real-life women reported reactions of unease and fright when confronted with the macabre. Such responses, in turn, raised questions about the fair sex's fortitude and suitability for consuming such displays.

Vigée-Lebrun's multifaceted reaction to the wax Venus may also have been at odds with her professional status as an accomplished female portrait painter. After all, her job detail required her to study the human form with a cultivated

68. On masculine strategies of veiling and unveiling women's reproductive secrets, see Jordanova, *Sexual Visions*, 87–110.

69. For an account of how excessive excitement incites debility, see Brown, *Elements of Medicine*.

70. On men finding the stench of cadavers debilitating, see Lister and Evelyn, *Voyage de Lister*, 69; and Sade, *Voyage d'Italie*, 64.

71. Détournelle, "Salon de Curtius," 20.

dispassion, not to mention sexual restraint. Painters like herself honed their skills by enrolling in any number of the anatomy courses geared at artists offered in the French capital. Such courses supplemented the knowledge of human anatomy that artists and trainees picked up in life classes and through the use of lay figures.[72] These new forms of training in anatomy were not primarily geared toward the fair sex, however, and aspiring female artists were often barred from attending life classes. Even the celebrated Vigée-Lebrun was restricted to drawing female nudes in those that she attended.[73]

Vigée-Lebrun's exclusion from drawing male nudes did not deter her from touring La Specola, where she examined waxworks of both sexes, while confronting her senses and probing the encounter in search of understanding. In this exercise, she transcended her femaleness and presented herself as a student of the Enlightenment. Her uncompromised curiosity and desire to reckon with the physical and emotional experiences of the wax museum incarnated the enlightened mantra "to know thyself."[74] Seen through this lens, Vigée-Lebrun offers her readers with a chance to vicariously witness a decidedly feminized heroic act, one in which she conquers revulsion and fear in the mature manner that Genlis imagined for her young charges. Her novel museum experience pointed to the transcendent value of sensibility and made a virtue of Vigée-Lebrun's willingness to parse and, indeed, "dissect" it.

Vigée-Lebrun's anatomical curiosity was more exemplary than excessive, and her femininity may even have elevated her contemplation of the wax woman. In her view, the act of confronting the Venus's entrails could not be reduced to mere medical-material instruction or philosophical inquiry. It symbolized a spiritual quest to make sense of the nature of the human body and its place in the universe. For her, experiencing the Venus on a soulful level was essential to reconciling the incongruency of its sheer materiality: its grotesque, coiled-up intestinal interior juxtaposed with its beautiful, smooth exterior.[75] Not long after her first visit to La Specola, she wrote: "It is not possible to contemplate the structure of the human body without feeling convinced of some divine power." She continued: "Despite what a few miserable philosophers have dared to say, in Mr. Fontana's laboratory one kneels and believes."[76] The wax Venus that haunted her like an unruly specter—or a praying skeleton, as depicted in

72. Anatomy courses for artists were advertised in the periodical press. See, e.g., Thiéry, *Le voyageur à Paris*, 201. On lay figures, see Monro, *Silent Partners*.

73. Oppenheimer, "'Charming Spectacle of a Cadaver,'" 76.

74. Cunningham, *Anatomist Anatomis'd*, 50.

75. On the changing relationship between the body's interior and exterior in anatomy, see Wagner, "Replicating Venus," 17–20.

76. Vigée-Lebrun, *Souvenirs*, 237.

William Cheselden's (1688–1752) *Osteographia* of 1733 (fig. 3)—also offered a glimpse of divine power. For another anatomy enthusiast, the marquise de Voyer d'Argenson (1734–83), the sight of cadavers prompted a similar sense of incredulity that left room for the divine. In a letter to her spouse about grappling with the unsightliness of corpses, Voyer d'Argenson stated that there was "nothing more singular in truth and more beautiful than this machine" and that "it is still incomprehensible that it could be a matter of chance."[77]

Women and the End-of-Century Wax Museum

Elite women's profound reverence for waxworks highlights the ennobling nature of anatomical inquiry, particularly when

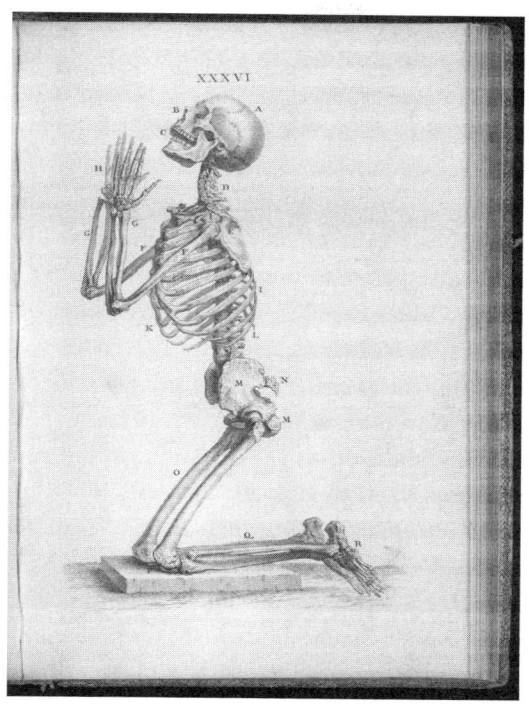

FIGURE 3 A praying skeleton in William Cheselden's *Osteographia; or, The Anatomy of the Bones* (1733), plate 36, elf WZ 260 C5240 1733. Photograph courtesy of Osler Library of the History of Medicine, McGill University.

guided by male oversight. Anatomy gave access to the profundity of the divine architect within the framework of Enlightenment sensibilities, which in the French context was also informed by the libertine-leaning ways of anticlerical deists.[78] This framework allowed for an enlightened individual to experience a natural visceral response to the interior of the human form while imbuing the encounter with the divine reverence it deserved. In this perspective, waxworks were easily linked to a mode of scientific rationalism that engendered permissible affective responses to the models, distinguishing them from the cruder waxwork traditions associated with entertainment culture.[79]

77. Delhaume, "Lettre 266, Paris, 28 février 1770," 405.
78. Hugues, "Esthétique et anatomie," esp. 156–58.
79. Jonathan Simon contends that waxworks within the eighteenth-century culture of sensibility were seen as primarily romantic rather than macabre objects ("Theater of Anatomy").

The boundary line drawn between vulgar and refined waxworks remained a thin one, however, and their geographic location was often a determining factor in how they were perceived. Marie-Marguerite Biheron's much-lauded waxworks thrived in Paris's historic Latin Quarter at the doorstep of medical and surgery students. In contrast, the waxworks enterprises of Philippe Curtius (1737–94, or Creutz/Curtz) were firmly situated in the capital's pleasure-seeking districts. From the 1770s Curtius showcased his exhibits in popular locales such as Boulevard du Temple, the Saint-Germain and Saint-Laurent fairs, and the seedy Palais-Royal.[80] While many of his waxworks were anatomically detailed and realistic, he was primarily viewed as a showman rather than a serious anatomist. The author of the *Chroniqueur désœuvré* lamented that Curtius's gallery catered to the misguided curiosity of clients seeking to purchase wax replicas of "small groups of fellows and libertines" to "decorate their boudoirs."[81] This characterization suggests, at best, that Curtius was providing nothing more than domestic appointments and, at worst, that he was pandering to less savory tastes.

Not all authorities appreciated the distinction often drawn between Biheron's anatomy cabinet and Curtius's waxworks. To them, the various poses of the Venus in well-known exhibits, no matter how dressed up as "science," were unilaterally sexually transgressive and thus warranted disapproval. Moreover, there was no agreement on the suitability of these waxworks for amateur female audiences, let alone for women requiring knowledge of the body for vocational ends. After all, if Vigée-Lebrun's visit to the wax museum allowed her to engage in on-the-job training, refining her craft almost came at the cost of her well-being.

By the end of the century, when Vigée-Lebrun made her pilgrimage to La Specola, the moral appropriateness of women engaging with waxwork models had become a growing concern. No consensus had been reached regarding a solution to the perceived problem. Charges that these models engendered sexual frisson were voiced by both men and women, who often shared in their disapproval of coeducation. Efforts to segregate tours of wax museums by sex were intermittently practiced throughout the century. For example, the title page of a sales catalog of Desnoües's models published in 1736 indicates that Friday was dedicated to ladies wishing to tour the waxworks now housed in London's the Strand. This arrangement, along with the exhibit's strong emphasis on models related to reproduction and childbirth, also appealed to midwives, who were

80. His waxworks were advertised in periodicals. See, e.g., La Mésangère, "Cire (*Figures* en)," 89; and Pujoulx, "Cabinet de figures." An inventory of the Saint-Laurent museum is provided in an insurance claim made after it was badly damaged by fire. See Archives Nationales, Z1J1694, "Visitte d'objets contentieux et état de degradation faits en une salle scituée a la foire, St Laurent et la à reqte de Sr Curtius . . . Locataire" signed Aug. 9, 1787, by Taboureur and Le Sage.

81. Mayeur de Saint-Paul, *Le chroniqueur désœuvré*, 136.

expected to exhibit exacting standards of modesty and decorum.[82] Expectations of female professionalism and discretion were not unique to French midwives, and in Bologna midwives undergoing training on the obstetrical collections at the Academy of Sciences (Accademia delle Scienze dell'Istituto di Bologna) were granted entry through a specially installed door, designed to safeguard their modesty.[83] This arrangement was at least less cumbersome than the practice in late sixteenth-century Montpellier of women concealing their faces behind masks while attending anatomy lessons at the university's Faculté de Médecine.[84]

The notion that aspiring midwives were expected to observe boundaries of decorum in training environments was nothing new in the eighteenth century, though the use of waxworks and other simulative models known as "phantoms" was.[85] Midwifery instructors of both sexes navigated the landscape of lifelike teaching tools and harnessed their novelty to attract students to their for-fee lessons. Midwives, man-midwives, and enterprising scientific showmen were all aware that the large and growing female consumer market for anatomical models demanded greater care and attention to codes of politeness. As the need for female instruction became more apparent, efforts to offer sex-segregated tours of anatomy spaces intensified. Women-only tours of museums provided by a female guide became the gold standard, breaking with the long-standing tradition of male tutelage of young women that was common in other forms of science education, such as chemistry.[86] Biheron held a distinct advantage over her male competitors as a woman: she could be trusted to teach those of her own sex with delicacy. This intergenerational female-to-female knowledge transfer was, of course, widely practiced in midwifery. Biheron's femaleness was also a selling point when she exhibited her waxwares in London in the early 1770s. A newspaper advertisement indicated that "ladies and persons of distinction may have other hours, giving timely notice."[87]

Objections to waxworks intensified as the eighteenth century wore on, and efforts to police anatomy teaching on lifelike models were renewed. A century after Desnoües's waxworks faced censure, obscenity laws were used to curb the use of models of female anatomy in medical teaching. In Philadelphia the American sex educator Frederick Hollick (1818–1900) faced charges for both the explicit content of his books and the "crime" of virtually dissecting a life-size papier-mâché female model in front of a mixed-sex audience.[88] Women also participated

82. Cock, *Catalogue of Several Curious Figures.*
83. Newman, *Fetal Visions*, 48.
84. Platter, *Journal of a Younger Brother*, 36–37.
85. Carlyle, "Phantoms in the Classroom."
86. Lehman, "Between Commerce and Philanthropy."
87. *Gazetteer and New Daily Advertiser.*
88. Haynes, "Obscenity, Sex Education," 166.

in moralizing screeds demanding sex-segregated tours of wax museums. When the Irish socialite and novelist Marguerite Gardner, Countess of Blessington (1789–1849), toured La Specola in 1830 with a female companion, she was scandalized to see young men and women "contemplating objects which, although highly useful for scientific purposes, are certainly of a character unfit for this promiscuous exhibition." She added that "knowing that we are fearfully and wonderfully created" does not mean that "we should witness the disgusting details of the animal economy in all its hideous and appalling nakedness and truth." Blessington promptly fled from the exhibit and doubled down on her belief that a museum was no place for female-male sociability; separate visits should be arranged.[89]

Female objectors like Blessington did not call for an outright ban of the fair sex's access to museums; rather, their primary moral concern was to ensure that encounters with waxworks occurred in the company of their own sex and without risking their reputations. Male tourists to Florence were also shocked by gender-mixed admission to the exhibition, but many of them proposed a solution contrary to Blessington's: women's exposure to nudity in any form, whether a newfangled waxwork or an ancient statue, irretrievably compromised their innocence and should be prohibited. In 1799, in the *Journal de Paris*, the painter and art critic Charles-Paul Landon (1760–1826) voiced his objection to young girls attending anatomy dissections or life classes. He believed that anatomy amphitheaters outfitted with instructional objects like skeletons were intended only for men, given that their degree of realism "leaves painful impressions on the soul" and "infallibly tarnishes" any pretense of modesty.[90]

The turn-of-the-nineteenth century backlash against women's free rein in anatomy amphitheaters or wax museums did not mean that there was no role for women in these spaces, either as visitors or as guides. If anything, this backlash presented the opportunity to erect a new moralizing discourse in which wax anatomies of an overtly pathological nature were used to provide young people of both sexes with an education aimed at curbing sex. Such was the aim of Jean-François Bertrand-Rival's (n.d.) wax cabinet, active from around 1775 to 1801 on the Rue Hautefeuille in Paris's Latin Quarter.[91] In 1798 he created a special display devoted to "objects concerning masturbation or onanism,"[92] including wax tableaux depicting the deadly consequences of self-stimulation. This was by all accounts a shocking spectacle that was intended to leave a lasting impression on the pubescent boys who toured the collection as part of school trips. One

89. Blessington, *Idler in Italy*, 215.
90. Landon, "Sur les femmes artistes," 844.
91. Le Minor, "Le cabinet des cires médicales."
92. Betrand-Rival, *Tableau historique*, 46–60.

figure of a boy in good health was juxtaposed with his alter ego fifteen months later, now "in a deplorable state by masturbation."[93] Girls were also warned off sexual activity through the model of a young woman with an ulcerated lung who had been "devoured by love."[94]

After visiting in 1800, the German writer Johann Georg Heinzmann (1757–1802) concluded that Bertrand-Rival's exhibition was a "remedy to ward off the corruption of Paris." He added that "here vice is entirely unmasked and exposed in all its natural nudity."[95] By 1807 the collection was described by the memoirist Louis-Marie Prudhomme (1752–1830) as a museum offering "a course in morality" in which visitors were exposed to the deleterious effects of onanism and venereal disease on human health, as well as the dangers of "impure women" on the moral integrity of the adolescent man. Bertrand-Rival's waxworks thus functioned primarily as a tool for moral and social education. His waxworks provided cautionary lessons, "a salutary advertisement for young people" in the Napoleonic age.[96]

Conclusion

The use of waxworks and other models by progressive educators, whether in the classroom, a hallway, or a museum, was part of the broader process of domesticating anatomy that unfolded during the eighteenth century. While gentlemanly science was associated with "gentle men," this did not exclude women from its study or, as we have seen in the case of Biheron at least, from its practice. A number of enlightened pedagogues found that some training was appropriate for the female mind, although most also concurred that provisioning elite girls with the skills needed to satisfy their future duties as wife and mother was the end game.

The increasingly utilitarian and pedagogical purposes that waxworks served in environments of learning help explain why midcentury audiences in Paris were permitted to indulge in them. The early days of policing wax cabinets in Paris, like Desnoües's, may have guided subsequent modelers in how to approach their own displays with an adequate level of moral restraint. The use of waxworks as teaching tools was nonetheless neither seamless nor universal, and it is evident that mixed-sex interactions with them generated criticism from the outset and continued to do so as the century wore on, whether from audiences, demonstrators, or curators. The shock-horror of anatomy for young girls and, indeed, women was a running theme in these critiques. The kinds of hallway anatomy

93. Betrand-Rival, *Tableau historique*, 47.
94. Betrand-Rival, *Tableau historique*, 25.
95. Heinzmann, *Mes matinées à Paris*, 367.
96. Prudhomme, *Miroir*, 258.

lessons that Genlis provided to conquer these fears were ultimately theoretical exercises that required translation into real-life experience by women like Vigée-Lebrun. On the flip side, the kinds of fantasies fueled by perverted female protagonists invented by male libertines voiced the larger social criticisms offered by conservatives, who feared women's unsupervised educational enfranchisement in anatomy.

The story of women's encounters with anatomical models is also a tale about the fundamental challenge of converting new enlightened educational schemes into practice and their uneven application in the real world of scientific learners. It is equally a narrative about the age-old fear of female bodies and the discordant views of them held simultaneously by thinkers of the age: women were perceived as lacking sexual self-control and moral restraint despite being held up as embodiments of all that is chaste and civilized. The numerous discourses around and the deeply contested nature of anatomical models in the age of Enlightenment in many ways mirror contemporary debates about what kinds of knowledge should be dispensed to school-age children about human sex and sexuality. It is clear that conservative-minded thinkers, then as now, fundamentally object to the liberatory power of sexual knowledge and self-knowledge.

MARGARET CARLYLE is assistant professor at the University of British Columbia Okanagan. She specializes in the history of medicine, science, and technology in early modern France and the Atlantic world. Her research centers on women's contributions to the development of medicoscientific knowledge, including their roles as experimentalists, inventors, artisans, and translators. She is completing two book projects, provisionally titled *Women and Anatomy in Enlightenment France*, which charts the role of women in the rise of modern anatomy, and *Delivering the Enlightenment*, which delves into the history of reproductive technologies in eighteenth-century France.

Acknowledgments

The author would like to thank the journal editors, special issue editors, peer reviewers, and copy editors for their invaluable input, which she hopes is reflected in the final version of this work.

References

Abramovici, Jean-Christophe. "The Comedy of Ignorance: Scenes of Sexual Initiation in Early Modern Pornographic Literature." In *Sex Education in Eighteenth-Century France*, edited by Shane Agin, 207–16. Voltaire Foundation, 2011.

Agin, Shane. "Sex Education in the Enlightened Nation." *Studies in Eighteenth-Century Culture* 37, no. 1 (2008): 67–87.

Arrêt de la Cour de Parlement du 19 août 1712 (Relatif aux démonstrations anatomiques faites par le sieur Desnoues, chirurgien, sur des figures en cire colorée). Paris, 1712.

Avant-coureur. Dec. 6, 1773.

Ballestriero, Roberta. "Anatomical Models and Wax Venuses: Art Masterpieces or Scientific Craft Works?" *Journal of Anatomy* 216, no. 2 (2010): 223–34.

Beaumont, Jeanne-Marie Leprince de. *Magasin des adolescentes, ou dialogues entre une sage gouvernante et plusieurs de ses élèves.* Vol. 1. Paris, 1797.

Benedict, Barbara M. *Curiosity: A Cultural History of Early Modern Inquiry.* University of Chicago Press, 2002.

Bertrand-Rival, Jean-François. *Tableau historique et moral des principaux objets en cire préparée et coloriée d'après nature qui composent le cabinet de J. Fs. Bertrand, ancien professeur de physiologie et d'accouchemens, et auteur dudit cabinet, rue Haute-feuille, no. 51.* Paris, 1798.

[Biheron]. *Anatomie artificielle.* Paris, 1761.

Blessington, Marguerite. *The Idler in Italy.* Paris, 1839.

Boulinier, Georges. "Une femme anatomiste au siècle des Lumières: Marie Marguerite Biheron (1719–1795)." *Histoire des sciences médicales* 35, no. 4 (2001): 411–23.

Brown, John. *Elements of Medicine.* Vol. 1. London, 1795.

Burger, Jacqueline. *Les maquettes de Madame de Genlis (1746–1830).* Musée des Arts et Métiers, 2000.

Carlyle, Margaret. "Artisans, Patrons, and Enlightenment: The Circulation of Anatomical Knowledge from Paris to St. Petersburg." In *Bodies Beyond Borders: Moving Anatomies, 1750–1950,* edited by Kaat Wils, Raf de Bont, and Sokhieng Au, 23–49. Leuven University Press, 2017.

Carlyle, Margaret. "Collecting the World in Her Boudoir: Women and Scientific Amateurism in Eighteenth-Century Paris." *Early Modern Women: An Interdisciplinary Journal* 11, no. 1 (2016): 149–61.

Carlyle, Margaret. "Phantoms in the Classroom: Midwifery Training in Enlightenment Europe." *KNOW: A Journal on the Formation of Knowledge* 2, no. 1 (2018): 111–36.

Carlyle, Margaret, and Katherine M. Reinhart, eds. "Anatomical Things." Special issue, *KNOW: A Journal on the Formation of Knowledge* 6, no. 1 (2022).

Casamaggi, Valerio Cantafio, and Armelle St-Martin. "Sade à Florence: Chronique et faits divers italiens dans *L'histoire de Juliette.*" *Dix-huitième siècle,* no. 41 (2009): 639–53.

Ceglia, Francesco Paolo de. "The Importance of Being Florentine: A Journey Around the World for Wax Anatomical Venuses." *Nuncius* 26, no. 1 (2011): 83–108.

Ceglia, Francesco Paolo de. "Rotten Corpses, a Disembowelled Woman, a Flayed Man: Images of the Body from the End of the Seventeenth to the Beginning of the Nineteenth Century; Florentine Wax Models in the Firsthand Accounts of Visitors." *Perspectives on Science* 14, no. 4 (2006): 417–56.

Chartier, Roger, Dominique Julia, and Marie-Madeleine Compère. *L'éducation en France du XVIe au XVIIIe siècle.* Société d'édition d'enseignement supérieur, 1976.

Cock, Christopher. *A Catalogue of Several Curious Figures of Human Anatomy in Wax, Taken from the Life. With Several Other Valuable Preparations (the Works of the Late Ingenious Mon. Denoue, of Paris, Who Was Forty Years in Making These Excellent Emblems Of Nature).* London, 1736.

Craske, Matthew. "'Unwholesome' and 'Pornographic': A Reassessment of the Place of Rackstrow's Museum in the Story of Eighteenth-Century Anatomical Collection and Exhibition." *Journal of the History of Collections* 23, no. 1 (2011): 75–99.

Cunningham, Andrew. *The Anatomist Anatomis'd: An Experimental Discipline in Enlightenment Europe.* Ashgate, 2010.

Dacome, Lucia. "Intimate Connections: Marie Marguerite Biheron and Her 'Little Boudoir.'" *Bulletin of the History of Medicine* 95, no. 3 (2021): 315–49.

Dacome, Lucia. *Malleable Anatomies: Models, Makers, and Material Culture in Eighteenth-Century Italy*. Oxford University Press, 2017.

Dacome, Lucia. "Women, Wax, and Anatomy in the 'Century of Things.'" *Renaissance Studies* 21, no. 4 (2007): 522–50.

Darnton, Robert. *The Forbidden Bestsellers of Pre-Revolutionary France*. W.W. Norton, 1995.

Daston, Lorraine, ed. *Things That Talk: Object Lessons from Art and Science*. Princeton University Press, 2004.

Delhaume, Sophie, ed. "Lettre 266, Paris, 28 février 1770." In vol. 1 of *Correspondance conjugale, 1760–1782: Une intimité aristocratique à la veille de la Révolution*, 404–6. Honoré Champion, 2019.

Delon, Michel. *L'invention du boudoir*. Grain d'Orage, 1999.

Desgenettes, René-Nicolas Dufriche. *Souvenirs de la fin du XVIIIe siècle et du commencement du XIXe, ou Mémoires*. Vol. 1. Paris, 1835.

Desnoües, Guillaume, and Domenico Guglielmini. *Lettres de G. Desnoues, professeur d'anatomie, et de chirurgie, de l'Académie de Bologne; et de Mr. Guglielmini, professeur de medecine et de mathematiques à Padoüe, de l'Académie Royale des Sciences et d'autres savans sur differentes nouvelles découvertes*. Rome, 1706.

Détournelle, Athanase. "Salon de Curtius." In *Aux armes et aux arts! Journal de la Société républicaine des arts*, 18–20. Paris, 1794.

Diderot, Denis. *Les bijoux indiscrets*. Paris, 1748.

Diderot, Denis. "Sur le même sujet la maison des jeunes filles." In vol. 10 of *Oeuvres complètes*, edited by Roger Lewinter, 612–17. Paris, 1771.

Didi-Huberman, Georges. *Ouvrir Vénus: Nudité, rêve, cruauté*. Gallimard, 1999.

Douglas, Aileen. "Popular Science and the Representation of Women: Fontenelle and After." *Eighteenth-Century Life* 18, no. 2 (1994): 1–4.

Dupaty, Louis-Marie-Charles-Henri Mercier. *Lettres sur l'Italie en 1785*. Vol. 1. Paris, 1788.

Ebenstein, Joanna. *The Anatomical Venus: Wax, God, Death, and the Ecstatic*. D.A.P. Publishers, Inc., 2016.

Etat de la médecine, chirurgie et pharmacie en Europe, et principalement en France. Par une Société de médecine. Paris, 1777.

Gargam, Adeline. "Between Scientific Investigation and Vanity Fair: Reflections on the Culture of Curiosity in Enlightenment France." In *Women and Curiosity in Early Modern England and France*, edited by Line Cottegnies, John Thompson, and Sandrine Parageau, translated by Matthew Hylands, 197–215. Brill, 2016.

Gargam, Adeline. "Marie-Marguerite Biheron et son cabinet d'anatomie: Une femme de science et une pedagogue." In *Femmes éducatrices au siècle des Lumières*, edited by Isabelle Brouard-Arends and Marie-Emmanuelle Plagnol-Diéval, 147–56. Presses universitaires de Rennes, 2007.

Gazetteer and New Daily Advertiser (London), Jan. 8, 1773.

Gelfand, Toby. "The 'Paris Manner' of Dissection: Student Anatomical Dissection in Early Eighteenth-Century Paris." *Bulletin of the History of Medicine* 46, no. 2 (1972): 99–130.

Genlis, Mme. de. *Adèle et Théodore ou lettres sur l'éducation contenant tous les principes relatifs à l'éducation des princes, des jeunes personnes et des hommes*. Vol. 1. Paris, 1782.

Genlis, Mme. de. *Adelaide and Theodore; or, Letters on Education (1783)*, edited and translated by Gillian Dow. Routledge, 2007.

Genlis, Mme. de. *Mémoires inédits de Madame la Comtesse de Genlis, pour servir à l'histoire des dix-huitième et dix-neuvième siècles.* Vol. 1. Paris, 1825.

Gill, Natasha. *Educational Philosophy in the French Enlightenment.* Farnham, 2010.

Goodden, Angelica. *Diderot and the Body.* Legenda, 2001.

Guerrini, Anita. "The Whiteness of Bones: Sceletopoeia and the Human Body in Early Modern Europe." *Bulletin of the History of Medicine* 96, no. 1 (2022): 34–70.

Guiffrey, J. "Le cabinet anatomique du chirurgien Desnoues." *Nouvelles archives de l'art français* 6, ser. 3 (1889): 163–64.

Halévi, Ran, ed. *Le savoir du prince du Moyen-Age aux Lumières.* Fayard, 2002.

Harvey, Karen. *Reading Sex in the Eighteenth Century: Bodies and Gender in English Erotic Culture.* Cambridge University Press, 2004.

Haynes, April. "Obscenity, Sex Education, and Medical Democracy in the Antebellum United States." In *American Sexual Histories*, edited by Elizabeth Reis, 165–86. 2nd ed. Wiley-Blackwell, 2012.

Heinzmann, Johann Georg. *Mes matinées à Paris: Voyage d'un Allemand à Paris et retour par la Suisse.* Paris, 1800.

Hilloowala, Rumy, and Joseph Renahan. "XVIII Century Anatomical Models at La Specola, Florence." *Anatomischer Anzeiger* 159, nos. 1–5 (1985): 141–58.

Hoffmann, Kathryn A. "Sleeping Beauties in the Fairground." *Early Popular Visual Culture* 4, no. 2 (2006): 139–59.

Huard, Pierre. "L'enseignement medico-chirurgical." In *Enseignement et diffusion des sciences en France au XVIII siècle*, edited by René Taton and Charles Bedel, 171–236. Hermann, 1964.

Hugues, Sylvie. "Esthétique et anatomie: Science, religion, sensation." *Dix-huitième siècle* no. 31 (1999): 141–58.

Jacob, Margaret C. "The Materialist World of Pornography." In *The Invention of Pornography: Obscenity and the Origins of Modernity, 1500–1800*, edited by Lynn Hunt, 157–202. Zone Books, 1996.

Jordanova, Ludmilla. *Sexual Visions: Images of Gender in Science and Medicine Between the Eighteenth and Twentieth Centuries.* University of Wisconsin Press, 1989.

Lalande, Joseph Jérôme de. *Voyage en Italie: Contenant l'histoire et les anecdotes les plus singulières.* Vol. 2. Paris, 1786.

La Mésangère, Pierre de. "Cire (Figures en)." In *Le voyageur à Paris: Tableau pittoresque et moral de cette capitale.* 2nd ed. Vol. 1. Paris, 1800.

Landon, Charles-Paul. "Sur les femmes artistes." *Le journal de Paris*, Mar. 31, 1799.

Ledoux. "Peinture—au citoyen Landon, peintre, auteur de plusieurs articles, sur les femmes artistes, insérés dans le Journal de Paris." *Journal des arts, de littérature et de commerce*, no. 1 (5 Thermidor year VII [July 23, 1799]): 1–3.

Lehman, Christine. "Between Commerce and Philanthropy: Chemistry Courses in Eighteenth-Century Paris." In *Science and Spectacle in the European Enlightenment*, edited by Bernadette Bensaude-Vincent and Christine Blondel, 103–16. Routledge, 2008.

Lehmann, Ann-Sophie. "Material Literacy." *Bauhaus Zeitschrift* no. 9 (2017): 20–27.

Le Minor, Jean-Marie. "Le cabinet des cires médicales du céroplasticien J. F. Bertrand à Paris (fin XVIIe–début XIXe siècle)." *Histoire des sciences médicales* 33, no. 3 (1999): 275–86.

Lemire, Michel. *Artistes et mortels.* Chabaud, 1990.

Lister, Martin, and John Evelyn. *Voyage de Lister à Paris en MDCXCVIII.* Translated by Société des Bibliophiles Français. Paris, 1873.

London Post and General Advertiser, May 31, 1736.

Lynn, Michael R. *Popular Science and Public Opinion in Eighteenth-Century France*. Manchester University Press, 2006.

Maerker, Anna. "Anatomizing the Trade: Designing and Marketing Anatomical Models in Medical Technologies, ca. 1700–1900." *Technology and Culture* 54, no. 3 (2013): 531–62.

Maerker, Anna. *Model Experts: Wax Anatomies and Enlightenment in Florence and Vienna, 1775–1815*. Manchester University Press, 2011.

Maerker, Anna. "Towards a Comparative History of Touch and Spaces of Display: The Body as Epistemic Object." In "Law and Conventions from a Historical Perspective." Special issue, edited by Rainer Diaz Bone, Claude Didry and Robert Salais. *Historical Social Research* 40, no. 1 (151): 284–300.

Mayeur de Saint-Paul, François-Marie. *Le chroniqueur désœuvré, ou l'espion du Boulevard du Temple*. 2nd ed. Paris, 1782.

Messbarger, Rebecca. *The Lady Anatomist: The Life and Work of Anna Morandi Manzolini*. Chicago University Press, 2010.

Messbarger, Rebecca. "The Re-Birth of Venus in Florence's Royal Museum of Physics and Natural History." *Journal of the History of Collections* 25, no. 2 (2012): 195–215.

Monro, Jane. *Silent Partners: Artist and Mannequin from Function to Fetish*. Yale University Press, 2014.

Muchembled, Robert. *Orgasm and the West: A History of Pleasure from the Sixteenth Century to the Present*. Translated by David Fernbach. Polity Press, 2008.

Naudin, Marie. "Stéphanie-Félicité, Comtesse de Genlis (1746–1830)." In *French Women Writers: A Bio-Bibliographical Source Book*, edited by Eva Martin Sartori and Dorothy Wynne Zimmerman, 178–87. Greenwood Press, 1991.

Newman, Karen. *Fetal Visions: Individualism, Science, Visuality*. Stanford University Press, 1997.

Oppenheimer, Margaret A. "'The Charming Spectacle of a Cadaver': Anatomical and Life Study by Women Artists in Paris, 1775–1815." *Nineteenth-Century Art Worldwide: A Journal of Nineteenth-Century Visual Culture* 6, no. 1 (2007): 59–85.

Panzanelli, Roberta, ed. *Ephemeral Bodies: Wax Sculpture and the Human Figure*. Getty Research Institute, 2008.

Pirson, Chloé. *Corps à corps: Les modèles anatomiques entre art et médecine*. Mare & Martin, 2009.

Platter, Thomas. *Journal of a Younger Brother: The Life of Thomas Platter as a Medical Student at Montpellier at the Close of the Sixteenth Century*. Translated by Seán Jennett, 36–37. F. Muller, 1963.

Porter, Roy. "Love, Sex, and Medicine: Nicolas Venette and his *Tableau de l'Amour Conjugal*." In *Erotica and the Enlightenment*, edited by Peter Wagner, 90–122. Peter Lang, 1991.

Prudhomme, Louis-Marie. *Miroir de l'ancien et du nouveau Paris*. Vol. 2. Paris, 1807.

Pujoulx, Jean-Baptiste. "Cabinet de figures." In *Paris à la fin du dix-huitième siècle*, 99–104. Paris, 1801.

Rattner Gelbart, Nina. *Minerva's French Sisters: Women of Science in Enlightenment France*. Yale University Press, 2021.

Riballier, Philibert, and Charlotte Catherine Cossor de La Cressonnière. *De l'éducation physique et morale des enfans des deux sexes*. Paris, 1785.

Riskin, Jessica. *Science in the Age of Sensibility: The Sentimental Empiricists of the French Enlightenment*. University of Chicago Press, 2002.

Sade, D. A. F. de. *Œuvres complètes*. Edited by Gilbert Lely. Vol. 16. Paris, 1973.

Sade, Marquis de. *Voyage d'Italie*. Edited by Maurice Lever. Fayard, 1995.

Saisselin, Rémy. "The Space of Seduction in the Eighteenth-Century French Novel and Architecture." *Studies on Voltaire and the Eighteenth Century* 319 (1994): 417–31.

Schnalke, Thomas. *Diseases in Wax: The History of the Medical Moulage*. Quintessence Publishing Company. 1995.

Simon, Jonathan. "The Theater of Anatomy: The Anatomical Preparations of Honoré Fragonard." *Eighteenth-Century Studies* 36, no. 1 (1995): 63–79.

Sonnet, Martine. *L'éducation des filles au temps des Lumières*. Éditions du Cerf, 1987.

Steinbrügge, Liselotte. *The Moral Sex: Woman's Nature in the French Enlightenment*. Translated by Pamela E. Selwyn. Oxford University Press, 1995.

Stephens, Elizabeth. *Anatomy as Spectacle: Public Exhibitions of the Body from 1700 to the Present*. Liverpool University Press, 2011.

Stephens, Elizabeth. "Venus in the Archive: Anatomical Waxworks of the Pregnant Body." *Australian Feminist Studies*, no. 64 (2010): 133–45.

St-Pierre, Armelle. "Sade's System of Perversity and Italian Medicine." In *Marquis de Sade and the Scientia and Techne of Eroticism*, edited by Frederick Burwick and Kathryn Tucker, 26–53. Cambridge Scholars Publishing, 2007.

Sutton, Geoffrey. *Science for a Polite Society: Gender, Culture, and the Demonstration of Enlightenment*. University of Colorado Press, 1995.

Terrall, Mary. "Gendered Spaces, Gendered Audiences: Inside and Outside the Paris Academy of Sciences." *Configurations* 3, no. 2 (1995): 207–32.

Terrall, Mary. "Natural Philosophy for Fashionable Readers." In *Books and the Sciences in History*, edited by Marina Frasca-Spada and Nicholas Jardine, 239–54. Cambridge University Press, 2000.

Thiéry, Luc-Vincent. *Le voyageur à Paris*. Vol. 1. Paris, 1788.

Tourneux, Maurice. *Diderot et Catherine II*. Paris, 1899.

Turner, J. G. "Sexual Awakening as Radical Enlightenment: Arousal and Ontogeny in Buffon and La Mettrie." In *Sex Education in Eighteenth-Century France*, edited by Shane Agin, 237–66. Voltaire Foundation, 2011.

Véron, Louis Désiré. *Mémoires d'un bourgeois de Paris*. Vol. 3. Paris, 1854.

Vigée-Lebrun, Elisabeth. *Souvenirs de Madame Vigée Le Brun*. Vol. 1. Paris, 1869.

Vila, Anne C. "'Ambiguous Beings': Marginality, Melancholy, and the Femme Savante." In *Women, Gender, and Enlightenment*, edited by Barbara Taylor and Sarah Knott, 53–69. Palgrave Macmillan, 2005.

Voltaire. *Candide and Other Stories*. Edited by Roger Pearson. Oxford University Press, 2006.

Wagner, Corinna. "Replicating Venus: Art, Anatomy, Wax Models, and Automata." *19: Interdisciplinary Studies in the Long Nineteenth Century* 24 (2017). https://doi.org/10.16995/ntn.783.

Walters, Alice N. "Conversation Pieces: Science and Politeness in Eighteenth-Century England." *History of Science* 35, no. 2 (1997): 121–54.

Exporting French Nature
The Transnational Origins of Taxidermy

YOTAM A. TSAL

ABSTRACT   In 1819 a monumental collection of natural history was transported from Paris to Edinburgh. In 1820, the manual written by its owner, Louis Dufresne, one of the first to use the term *taxidermy*, was translated from French into English for the use of museums and travelers in the Anglophone world. Archival research and computational methods make it possible to chart the transportation of the collection and the translation and reception of the manual. This episode in the exportation of a highly refined, perfected, and rulebound version of the natural world gave the French a competitive advantage in the early nineteenth-century transnational market for natural history knowledge and specimens.

KEYWORDS   natural history, zoology, museums, transnational science, France

In 1819 Louis Dufresne, the chief taxidermist at the French National Museum of Natural History in Paris, sold his private natural history collection to the University of Edinburgh for an impressive $160,000 in today's value.[1] Dufresne and his wife, whose name remains unknown in the historical record, amassed the collection together. In the standards of the period, Dufresne's collection was considered both expansive and variegated. It boasted over eighteen thousand specimens of insects, birds, shells, fossils, and other aquatic organisms from places as diverse as Australia, southern Africa, and the Americas.[2]

The Scottish university purchased Dufresne's collection to form part of the inaugural collection of specimens at its newly established museum of natural history, one that locals hoped "[would] become one of the finest in Europe."[3]

---

1. This estimation was made with the help of the British National Archives' online currency converter services (https://www.nationalarchives.gov.uk/currency/#currency-result).

2. For the number of specimens, see *Catalogue des collections d'objets d'histoire naturelle formant le cabinet de M. L. Dufresne, naturaliste au Jardin du Roi* (Paris, 1815). The catalog is kept at the National Museums Scotland Collection Centre, Edinburgh. See also Thomas Brown to Robert Jameson, Mar. 29, 1819, Gen. 1801/2 / Coll-198, no. 16, Centre for Research Collections, Edinburgh University Library (hereafter CRCEUL), Scotland; and *Scots Magazine and Edinburgh Literary Miscellany*, 361.

3. *Scots Magazine and Edinburgh Literary Miscellany*, 171. See also Brown to Jameson, Mar. 29, 1819; and a letter addressed to Robert Jameson, July 10, 1819, Gen. 1801/2/ Coll-198, no. 59, CRCEUL.

French Historical Studies • Vol. 48, No. 2 (May 2025) • DOI 10.1215/00161071-11626737

FIGURE 1 Insects in Dufresne's collection, kept at the National Museums Scotland Collection Centre, Edinburgh. Photograph by the author.

One Scottish commentator referred to the Scottish collection after the French acquisition as "a splendid monument of national glory."[4] According to another observer, when appending the French purchase to the university's existing inventory of specimens, the collection of birds alone amounted to about three thousand specimens, making it "the most extensive in Great Britain, and not exceeded by many on the continent" (fig. 1).[5]

Following the arrival of Dufresne's collection on Scottish shores, the explorer and naturalist Sarah Bowdich published an English translation of Dufresne's taxidermy manual, which he had composed in 1803 as an entry for a French dictionary dedicated to natural history and its applications in agriculture, which later appeared in several editions.[6] The English version, *Taxidermy; or, the Art*

4. Miller, *Popular Philosophy*, 61.
5. Stark, *Picture of Edinburgh*, 182.
6. *Taxidermy; or, The Art of Collecting, Preparing, and Mounting Objects of Natural History*. The word *taxidermy* derives from the Greek words *taxis*, "arrangement," and *derma*, "skin." Thus it means literally the arrangement of skin.

of Collecting, Preparing, and Mounting Objects of Natural History, became highly popular in the Anglophone world, as evidenced by the fact that it was reprinted five times.[7]

This transnational episode in the history of preservation is particularly significant for two reasons. First, the newly French term *taxidermy*, which Dufresne popularized, became a global standard for naming preservation practices. Second, the preservation technology based on arsenic soap that was used at the time at the Parisian museum, and that Dufresne publicized, was subsequently replicated worldwide until the 1980s.[8]

I situate the emergence of modern taxidermy within the transnational context of France's natural history "Golden Age" in the early nineteenth century.[9] Extending Paula Findlen's observation that nature could be seen as a form of patrimony worthy of preservation, I ask, What happens when a natural patrimony of one nation becomes a natural patrimony of another?[10] I argue that the sale and purchase of Dufresne's collection illuminate France's competitive advantage in the burgeoning transnational market for natural history specimens and know-how that accompanied scientific investigation.[11] The competitive advantage of the French National Museum of Natural History was rooted in the quality and variety of its specimens, the museum's flexible transnational networks, and Paris's symbolic prowess in the popular imagination of the Anglophone world.

. . .

Recent scholarship on taxidermy has stressed that the preserved specimen is as much a cultural artifact as it is a relic of nature.[12] As Alan S. Ross notes, "The very term 'preservation' is in fact misleading: an elaborate and deliberately engineered process of transformation had been performed on the raw material of the dead animal body."[13] The preserved specimen is thus both a

7. Orr, "Stuff of Translation," 27. So far the transportation of the collection and the translation of the manual have been investigated separately. See Orr, "Stuff of Translation "; and Sweet, " Collection of Louis Dufresne."

8. On the trajectory of the use of arsenic soap, see Jones, "Rhinoceros and the Chatham Railway," 714; and Marte et al., "Arsenic in Taxidermy Collections," 144.

9. On France's scientific "Golden Age," see Spary, *Utopia's Garden*, 2. In literature about preservation, Dufresne's use of arsenic soap and use of the term *taxidermy* are often mentioned, albeit in passing, as a pivotal moment in the inception of modern taxidermy. See, e.g., Habermeyer, "Skin Walking," 321.

10. Findlen, *Possessing Nature*, 397.

11. For the place of French naturalists in European science, see Crosland, "Science Empire in Napoleonic France," 29.

12. On the history of taxidermy, see Farber, "Development of Taxidermy "; and Wonders, *Habitat Dioramas*. For the new wave of scholarship on the matter, see Ross, "Preserving the Animal Body "; Alberti, *Afterlives of Animals*; Aloi, *Speculative Taxidermy* ; Patchett, "Putting Animals on Display "; Péquignot, "Histoire de la taxidermie en France "; and Poliquin, *Breathless Zoo*.

13. Ross, "Animal Body as Medium," 87–88.

(decontextualized) natural thing and a cultural artifact as any other human-made object, occupying a liminal space as simultaneously natural and artificial. The attentiveness to the relationship between the artificial and natural dimensions of natural historical specimens was especially prominent in the French context, wherein naturalists promoted what they perceived as a "perfected" version of nature.[14]

Although taxidermy has attracted growing scholarly attention, historians have largely overlooked the transnational dimensions of preservation history and the inherent malleability of preserved specimens. These specimens demand continual maintenance, including adjustments in collections, repairs, transportation, and exchanges across various locations. The preserved specimen, I argue, was an entity in motion not only between nature and culture but also within culture and between cultures. I intervene in the scholarship on taxidermy by examining the movement and knowledge dissemination of the preserved body across national boundaries. I contribute to the scholarship on taxidermy by bridging the perspectives of material culture and the history of transnational science.[15] The transnational context of material culture is vital in the case of natural commodities, from sugar and indigo to shells, insects, and birds. In this context, the preserved body was in motion in more dimensions than one. The transnational circulation of these goods across Europe constituted an extension of the transportation of colonial specimens to European cities. The history that this article provides is thus one of imperial science as much as it is of transnational science.

A transnational perspective is essential for understanding French natural history in general and the production of specimens within the Hexagon in particular. Recent studies of eighteenth- and early nineteenth-century natural history have shown that the French community of specimen producers was deeply interconnected with cosmopolitan networks of naturalists. The same forces often promoted both scientific nationalism and cosmopolitanism.[16] Although France did not possess a large colonial empire during this period, it witnessed three decades of scientific expeditions, as well as an extensive mobilization of personnel, knowledge, and specimens from across the country and the systematic acquisition of specimens from the rest of Europe. Consequently, Paris became a major hub in the European transnational market for natural specimens, and French naturalists

14. For an exploration of how French naturalists used the representation of the natural world as a vehicle for the transformation of society, see Spary, *Utopia's Garden*.

15. Emma Spary argues that "when the history of natural history is rewritten from the standpoint of material culture, a different story needs to be told" ("On the Ironic Specimen," 1034). For more on the perspective of material culture in historical research, see Findlen, *Early Modern Things*. See also Bourguet, *Instruments*.

16. See, e.g., Easterby-Smith, *Cultivating Commerce*; and Parsons and Murphy, "Ecosystems Under Sail."

took pride in being custodians of nature on a global scale.[17] Thus, Dufresne operated in a world that was simultaneously cosmopolitan and influenced by strong national sentiments.

Paris was also an epicenter for gathering and disseminating knowledge about preservation. Preservation techniques did not develop in a linear process. The method using arsenic soap that Dufresne promoted was invented in the mid-eighteenth century in a small provincial town, long before the first edition of his manual appeared in 1803. It was only during the French Revolution, when state funding and centralization of scientific knowledge intensified, that this local technology was disseminated to the national capital and subsequently diffused to the entire nation.[18] Seventeen years later, on the heels of the centralization of European and colonial scientific knowledge in Paris, Dufresne exported this technology to the Anglophone world via the translation of the manual and the transportation of the collection. This article focuses on how this exchange took place.

In what follows, I examine the cross-border circulation of French knowledge. First, I analyze the collection's catalog as a window into the collection's global reach and value for scientific and commercial purposes. Second, using network analysis, I map out the Parisian museum's role as a transnational hub of information and artifact exchange that transcended its formal institutional boundaries. Third, I trace the reception of the manual and the collection in the Anglophone world and its broader sociocultural contexts. I conclude the article by highlighting the role of a private collection in a transnational market of natural history specimens.

A "Classical" Collection

Dufresne's collection was a sought-after commodity in the British natural history market.[19] By 1818 representatives from two preeminent British institutions, the University of Edinburgh and the British Museum in London, expressed an interest in purchasing the collection.[20] Why was Dufresne's collection valued so

17. Historians of science have shown that it was almost self-evident to people in eighteenth-century France for their immediate surroundings to serve as a universal standard. See, e.g., Van Damme, *Paris, capitale philosophique*; Alder, *Measure of All Things*; and Tsal, "Newsworthy Nature."

18. The recipe using arsenic soap was invented by Jean-Baptiste Bécoeur in the mid-eighteenth century. Bécoeur kept it a secret until his death. See Rookmaaker et el., "Ornithological Cabinet." The recipe was first published by François-Marie Daudin in *Traité élémentaire et complet d'ornithologie*, 445; and by Pierre François Nicolas in *Méthode de préparer et conserver les animaux*, 28. Arsenic soap was used at the time also by several Anglophone naturalists. See Farber, "Development of Taxidermy," 561n42.

19. According to later reports, the emperors of Austria and Russia also made offers to buy Dufresne's collection in 1819. See Chambers, *Biographical Dictionary of Eminent Scotsmen*, 385.

20. See William Leach to Robert Jameson, Nov. 15, 1818, Gen. 1801/2 / Coll-198, no. 4, CRCEUL. According to a later account, the trustees of the British Museum decided not to buy the collection. See Swainson, *Taxidermy*, 74.

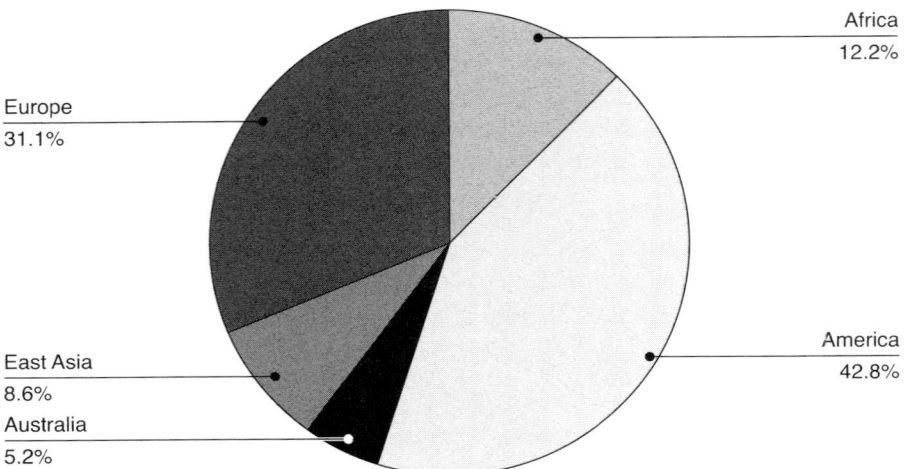

Europe
31.1%

Africa
12.2%

East Asia
8.6%

Australia
5.2%

America
42.8%

FIGURE 2 Origins of specimens.

highly by contemporaries? Part of its appeal can be found in its contents. A catalog of Dufresne's collection from 1815 provides an aperture into the scope of the collection.[21] Of all the specimens in the collection, the catalog displays the most elaborate information on birds, encompassing facts concerning the origin locality of a bird specimen; whether a specimen is male, female, young, or a rare species; the season it was caught; and a species classificatory category. I used the catalog of the bird collection as a window into Dufresne's collection as a whole.[22]

The first strength of the collection is its global reach. The catalog mentions 1,305 bird specimens.[23] The existing data suggest that Dufresne's collection was largely made up of "exotic" specimens. From the 898 specimens of which origins are mentioned, only 279 are European (see fig. 2). The largest group of specimens (385) came from the New World.[24] However, a sizable number of specimens arrived from Africa (110), East Asia (77), and Australia (47). Except for one African specimen that Dufresne noted having received from the French traveler François Levaillant, we do not know who provided which specimens.[25]

21. See *Catalogue des collections d'objets d'histoire naturelle*. There is also a later catalog: Dufresne, *Catalogue des collections d'histoire naturelle*.

22. On the origins of the British Museum, see Delbourgo, *Collecting the World*. On the French National Museum of Natural History, see Laissus, *Le Muséum national d'histoire naturelle*; and Spary, *Utopia's Garden*. See also Lacour, *La république naturaliste*.

23. A summary of the catalog from 1818 mentions fifteen hundred specimens, including one hundred duplicates. See William Leach to Patrick Neill, Mar. 1, 1818, Gen. 1801/2 / Coll-198, no. 82, CRCEUL.

24. Thus, together with the specimens from Europe (279) and Africa (110), the Atlantic World dominated the collection (774 specimens).

25. That one specimen was the species Martin pêcheur d'Afrique.

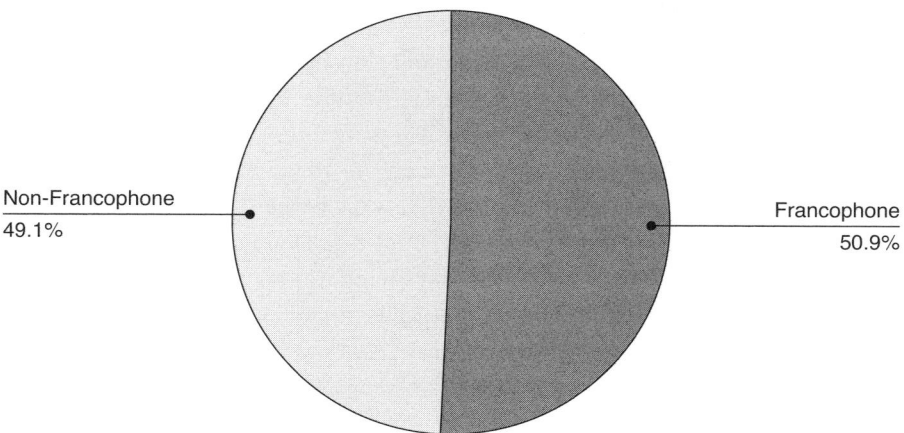

Non-Francophone
49.1%

Francophone
50.9%

FIGURE 3 Francophone versus non-Francophone origins of specimens.

However, from the existing information on the origins of specimens in the official collection of birds in the Parisian museum, one can estimate the sources of
the specimens.[26] Some of these specimens were given to the museum by French
travelers and colonists. Indeed, from the data I amassed on Dufresne's bird collection, 296 specimens originated in localities with French colonies or areas of
influence during the four decades when Dufresne amassed his collection. The
source of almost half of these specimens (and the locality with the most specimens
in general) is French Guyana (214), France's gateway colony in South America.
Senegal (22 specimens), and Haiti (10) are also well represented in the collection.
Thus, together with the 172 specimens that originated in France, the Francophone part of the collection consisted of 457 specimens (fig. 3).[27]

This number of 457 specimens suggests that roughly half of the bird specimens in Dufresne's collection whose origins are known did *not* originate in
French areas of influence.[28] Among the specific localities Dufresne mentioned,
beyond general areas such as "North America," are "Bengale" (27 specimens),
the Cape of Good Hope (20), Puerto Rico (19), Brazil (19), and "Carolina" (7). The
high number of specimens from outside France's areas of influence demonstrates
that the collection's origins were not simply confined to the French Empire but were

26. See Jansen, "Towards the Resolution of Long-Standing Issues."

27. This number includes sixteen specimens from Louisiana and Canada that were not French colonies in Dufresne's time but remained Francophone.

28. Or at least these specimens were not clearly mentioned as such. Many of them are mentioned as
originating from general areas, such as "Africa" and "North America." I treated those specimens as originating in areas outside French influence since, given the precision in mentioning French areas of influence
("Cayenne," "Saint Domingue," etc.), it is likely that those that have a more general indication of place were
procured outside the Francophone world.

global in scope. More fundamental, the French naturalist community could perceive it as French precisely because it was global, and global because it was French. As historian Elise Lipkowitz shows, during the Revolution and the Napoleonic Wars French naturalists and government officials justified the seizure of collections of other European nations as a cosmopolitan act made on behalf of humanity as a whole.[29] In an era that historian David Bell has called the "first total war," France acquired a wealth of specimens not only through leveraging its own scientific networks but also through appropriating specimens belonging to other European nations.[30] Thus, in the early nineteenth century owning a "quintessentially French" collection of natural history meant that one could possess specimens belonging to other European nations. The French collection was a secondhand collection, and even a third-hand one, because before becoming the property of these other European nations, these specimens were often forcibly taken or bought from Indigenous peoples in European colonies. In fact, Dufresne received sixty-four South American specimens from the confiscated "Lisbon Cabinet," which arrived at the Parisian museum in 1808.[31]

Scientific expeditions are probably another source of some of Dufresne's specimens. A notable one is the Baudin expedition of 1800–1803, approved by Napoleon, with the intent to map out the coast of New Holland (now Australia). Through this expedition, a large number of specimens were introduced to the national museum in Paris.[32] Specimens acquired during a series of expeditions from the 1790s to 1815 might explain the existence of specimens from such places as Brazil and Puerto Rico in Dufresne's collection. We also know that the museum in Paris exchanged specimens with institutions in other countries. This was the case, for example, in 1807, when the Parisian museum received specimens from the natural history museum in Vienna, and in the 1790s, when several specimens were acquired from the British collectors Joseph Banks and William Bullock.[33]

Thus the global reach of the Parisian museum—and of Dufresne's private collection, which was amassed in its shadow—was a result not merely of colonial control but also of the French state's ability to procure specimens in the inter-imperial and transnational arenas via expeditions, confiscations, and exchanges.[34]

---

29. Lipkowitz, "Seized Natural-History Collections," 25. On the question of science and nationalism in the period, see Daston, "Nationalism and Scientific Neutrality Under Napoleon."

30. Bell, *First Total War*. On the looting of German art collections by the French in the period, see Goff, *God Behind the Marble*.

31. Jansen, "Bird Collection," 90.

32. Jansen, "Towards the Resolution of Long-Standing Issues."

33. Jansen, "Bird Collection," 85, 90.

34. On the relationship between science and state in France in the period, see Chappey, "Enjeux sociaux et politiques"; Fox, *Savant and the State*; Gascoigne, *Science and the State*; Gillispie, *Science and Polity in France*; and Osborne, "Applied Natural History and Utilitarian Ideals."

Beyond its global dimensions, Scottish and French observers considered the pristine condition and scientific value of Dufresne's collection desirable. Thanks to the use of the arsenic soap recipe, Dufresne's specimens were well preserved. As historian Dorinda Outram puts it, museum officials such as Dufresne were "paid by the state to represent the practice and standards of the inner scientific community."[35] Dufresne was not just responsible for maintaining such practices and standards; as the first man to hold his position, he was tasked with creating them. His taxidermy manual, his manufacturing of specimens, and the training of personnel under his direction were intended to develop a universal standard for preservation. Through the prism of the meter's invention, historian Ken Alder has shown that the French Revolution gave savants the opportunity to implement their vision of a rational and coherent system of measurement that "would induce its users to think about the world in a rational and coherent way."[36] Likewise, Dufresne sought to rationalize "best practices" in collecting and preservation. In the aforementioned dictionary entry, for example, Dufresne stated:

> The employees of the zoology laboratory at the Museum, on their arrival in this establishment, were in the habit of passing the iron before stuffing the neck; all have given up this habit, and have preferred that which we have just indicated.[37] More than two thousand birds, among the number of those which now adorn the galleries of the Museum, have been mounted in this manner, which proves quite obviously that our method is not impracticable.[38]

He also mentioned proudly in a later edition that, since the publication of the first edition, "The number of people engaged in taxidermy has significantly increased, not only in Europe, but in all parts of the world, and we have had the satisfaction of observing that almost all the animals sent to the Museum of Paris, are prepared in the manner that we have outlined."[39] Indeed, Dufresne's specimens won accolades from the Scottish press and British observers attuned to the cross-border

35. Outram, *Georges Cuvier*, 119.
36. Alder, *Measure of All Things*, 2.
37. The term *iron* refers here to the iron wire used to hold the specimen from within.
38. Dufresne, "Taxidermie," 540. The entry was published in 1820 as a separate publication, titled *Taxidermie*: "Les employés du laboratoire de zoologie au Muséum, à leur arrivée dans cet établissement, étaient dans l'usage de passer le fer avant de bourrer le cou; tous ont renoncé à cette habitude, et ont préféré celle que nous venons d'indiquer. Plus de deux mille oiseaux, dans le nombre de ceux qui ornent maintenant les galeries du Muséum, ont été montés de cette manière; ce qui prouve assez évidemment que notre méthode n'est pas impraticable."
39. Dufresne, "Taxidermie" (1819), 523: "Le nombre des personnes qui s'occupent de *taxidermie* s'est singulièrement accru, non-seulement en Europe, mais dans toutes les parties du monde, et nous avons eu la satisfaction d'observer que presque tous les animaux envoyés au Muséum de Paris, sont préparés d'après les procédés que nous avons indiqués."

exchange.[40] Scottish periodicals touted Dufresne's exemplary preservation of the specimens. One commentator stated that the collection "arrived in excellent condition" and that the bird specimens in it were "in a state of perfect preservation."[41] Another observer praised Dufresne's collection, informing the readers that "the birds are preserved in the same manner as they were at the department at the Museum of the Jardin du Roi in Paris, to which grand institution M. Dufresne is chief naturalist." These specimens, they added, were "in general very fine, as M. Dufresne availed himself of every opportunity of procuring the best, to supersede those which were defective in his cabinet."[42]

Another aspect of the collection that made it valuable for scientific research was the variety of its specimens. As historian Lorraine Daston argues in another context, in the eighteenth and early nineteenth centuries "naturalists worthy of the name based their species descriptions on as wide a range of specimens as possible."[43] As the head of the laboratory in the Parisian museum, Dufresne had access to a massive volume of specimens flowing in and out of the institution. His private collection contained 100 duplicate bird specimens.[44] It boasted no fewer than 152 "rare" bird specimens, 62 of which were described as "very rare," "extremely rare," or "unique." Dufresne described 23 of the specimens as "new," "undetermined," or "unpublished," and the collection even included two new genera. The collection also contained 24 varieties of species. The access to a large volume of specimens was particularly important in the case of a temporal variety of specimens. Dufresne's collection contained many specimens that represented different stages in the life of a species. Overall the collection included 224 specimens identified as female, 271 identified as male, 120 as in a certain age group, and 15 specimens caught in different seasons. To one Scottish observer, the variety of Dufresne's collection was "highly valuable for the purposes of study, as the change of plumage from the young to the adult, and the difference between the male and the female, are the most perplexing circumstances in the study of natural history."[45]

The arrangement of specimens in the Dufresne's collection was another scientific contribution. As a Scottish observer noted, the specimens were "so put

40. The English naturalist William Swainson, for example, wrote in 1840 that the museum of the university, which was "principally composed of the well-known and valuable collection of M. Dufresne," was noteworthy because it "excites the admiration of all who have visited it, for the beauty and perfection of the specimens, and the neat manner of their arrangement" (*Taxidermy*, 74).

41. *Scots Magazine and Edinburgh Literary Miscellany*, 361.

42. *Scots Magazine and Edinburgh Literary Miscellany*, 170.

43. Daston, "Type Specimens and Scientific Memory," 166. On type specimens in zoology, see Farber, "Type-Concept in Zoology."

44. See Leach to Neill, Mar. 1, 1818.

45. *Scots Magazine and Edinburgh Literary Miscellany*, 170.

up as to be capable of any arrangement the Professor of Natural History may choose to adopt, and besides, are admirably fitted for the purpose of study." The use of cutting-edge Parisian classification systems was a valuable attribute. The eight hundred eggs in the collection were "accurately named," and the nearly four thousand shells were "arranged and named according to the system of Lamarck."[46] In a letter addressed to a colleague, a representative of the University of Edinburgh stressed the classification of the specimens as an important rationale for the acquisition. He underlined the range of genera and species and the system of naming and arrangement of the specimens as conducive to expanding the collection without compromising the classification of the entire collection.[47] In his efforts to secure funding for the purchase, the university official further noted that the specimens were classified in accordance with specimens in the Parisian museum and with those of the celebrity French ornithologist and traveler François Levaillant.[48]

Within the framework of postrevolutionary France's preoccupation with universalizing science, Dufresne's collection was recognized as a "classic" by many contemporaries due to its global scope and contributions to scientific research. For example, Dufresne provided his Scottish counterparts with a testimonial from the French museum's renowned professors as endorsement for his private collection. In a joint statement, Etienne Geoffroy Saint-Hilaire, Georges Cuvier, and Jean-Baptiste Pierre Antoine de Monet, chevalier de Lamarck, concluded that the superior quality of the collection, Dufresne's expertise in taxidermy, his access to a vast global network of travelers, and the general quality of his collections would be an asset. They asserted that the collections "have become a sort of classical authority" and could "in their present state form a National Museum" in Edinburgh.[49] Scottish observers likewise embraced this view. One referred to the new acquisition as "the classical and magnificent collection of birds purchased by the college from M. Dufresne of Paris."[50] Another designated it a "classical collection of zoology, purchased by the University of Edinburgh, from M. Dufresne of Paris."[51]

Comparison with a contemporary collection that the University of Edinburgh also considered for purchase provides insight into the comparative advantage of Dufresne's collection in the transnational scientific market, even compared

46. *Scots Magazine and Edinburgh Literary Miscellany*, 361. On the classification systems of the museum, see Deleuze, *Histoire et description*, 435, 617.

47. See Leach to Neill, Mar. 1, 1818.

48. See Leach to Neill, Mar. 1, 1818.

49. See certificate from the staff of the National Museum of Natural History, Gen. 1801/2/Coll-198, no. 5.

50. Stark, *Picture of Edinburgh*, 182.

51. Miller, *Popular Philosophy*, 60.

to a collection from London, despite its geographic proximity. The collector William Bullock marketed the sale of his collection to the university by emphasizing its larger size.[52] He critiqued his rival's collection by stating that "Mr. Dufresne's collection, though good and cheap, will in your noble rooms be only a Lilliputian in a Brobdingnagian Castle."[53] Eventually, the university elected not to purchase Bullock's collection, mainly for "want of room."[54] Differences in cost and quality appeared to play an equally significant role in the university's preference for Dufresne's collection. In comparison to the cost of Dufresne's collection, which was sold for 2,240 guineas, Bullock's priced his collection at no less than 9,000 guineas.[55] Furthermore, while university representatives complained about the poor quality of some of Bullock's specimens, they praised the pristine condition of those in Dufresne's collection.[56] While Bullock's collection was established for the purposes of entertainment as much as science, Dufresne's collection was all about science, and as such it was an exemplar of the postrevolutionary French state's centralization of power.

From Private to National

The transportation of the collection involved adapting a private French cabinet of natural history to the needs of a burgeoning British public institution. This adaptation necessitated intermediaries capable of performing the "alchemy" of transformation. Additionally, it required a physical environment conducive to such transformation, one that had already undergone transnational exchanges and had sufficient flexibility to accommodate unexpected challenges and contingencies smoothly.

The staff of the Parisian national museum promoted the institution as a space that specialized precisely in this type of "alchemy." With their vast network of informants across the world, the expeditions they supported, and the many formerly foreign and privately owned specimens in their institution, its professors prided themselves for creating universal knowledge from private and foreign sources of data. For Parisians, their museum was nothing less than "the unique point of the globe where almost the universality of the beings of nature

52. Sweet, "William Bullock's Collection," 23.
53. See William Bullock to Robert Jameson, Apr. 3, 1819, letter B4, quoted in Sweet, "William Bullock's Collection," 24.
54. See William Leach to Robert Jameson, Apr. 14, 1819, letter B5, quoted in Sweet, "William Bullock's Collection," 24–25. The university did, however, purchase individual specimens from the Bullock collection. See, e.g., *Edinburgh Magazine and Literary Miscellany*, vol. 83, 1819, 555.
55. Brown to Robert Jameson, Mar. 29, 1819, Gen. 1301/2 / Coll-198, no. 16, CRCEUL.
56. On the complaints about Bullock's specimens, see Sweet, "William Bullock's Collection," 30. For a praise of Dufresne's specimens, see Brown to Jameson, Mar. 29, 1819.

is reflected," as one observer put it.[57] The museum's staff reorganized and reclassified collections, displayed them in new contexts, and even disseminated specimens it produced to other institutions in the Hexagon and across the world.[58] Dufresne, in fact, had taken part in the transnational exchange of specimens, particularly with British collaborators, prior to the sale of his collection to Edinburgh.[59] With the reopening of the museum's gates to foreigners in the wake of Napoleon's defeat in 1814, the institution became an even more conducive space for managing a transnational operation of this scope.[60]

Historians trace a long historical trajectory of transnational scientific standardization, from the emulative practices of the early modern republic of letters to the formal conventions of the competitive international scientific arena of the late nineteenth century.[61] Dufresne's case illustrates the extent to which the early nineteenth century was a transitional period characterized by efforts to adapt knowledge across national boundaries. The absence of comprehensive institutional infrastructures and standardized scientific exchanges between nations, coupled with the increasing availability of resources—particularly those accessible through the National Museum of Natural History—created fertile ground for scientific entrepreneurship. This entrepreneurship involved active participation of both private and public actors, each of which stood to benefit significantly from the sale of such a prestigious collection.

An examination of the network of actors involved in the transportation of the collection reveals the extent to which the transnational circulation of knowledge during this period depended on loose ties between private and governmental actors across national boundaries.[62] The flexibility of these ties facilitated the exportation of knowledge while enabling adaptations to the new environment.[63]

57. Gérardin, *Tableau élémentaire d'ornithologie*, 1: "Le Muséum de Paris est le point unique du globe où la presque universalité des êtres de la nature se réfléchit."

58. When the confiscated Dutch Stadholder's collection of natural history arrived in Paris in 1795, for example, the Parisian museum's professors reclassified the specimens according to the systems of Buffon and Linnaeus that were used in Paris. See Hendriksen, "Animal Bodies Between Wonder and Natural History," 1125.

59. *Annales du Muséum d'histoire naturelle*, 178–80, 464. See also Sweet, "Collection of Louis Dufresne," 38–39.

60. Orr, "Women Peers in the Scientific Realm," 41.

61. See Alder, "Scientific Conventions," 23, 34. On eighteenth-century transnational natural history, see Easterby-Smith, *Cultivating Commerce*. On international science in the late nineteenth and twentieth centuries, see Crawford, "Universe of International Science"; Crawford et al., *Denationalizing Science*; and Heilbron, *Dilemmas of an Upright Man*.

62. As a growing number of historical studies have shown, computational network analysis is invaluable in understanding patterns in scientific and literary exchange within and beyond national boundaries. See Langmead et al. "Towards Interoperable Network Ontologies."

63. The network includes all the known actors involved in the transfer of the collection and those involved in the translation of Dufresne's manual, encompassing not only individuals who corresponded between Paris and Edinburgh but also those mentioned in these letters. The network also includes actors

The Parisian museum served as a major hub for these transnational connections and the required adjustments.[64]

The transfer of the Dufresne collection was a complex operation involving multiple stages: packing, cataloging, unpacking, corresponding, shipping, transferring funds, and navigating the transnational aspects, such as customs duties. This process engaged university and museum officials, diplomats, the British navy, and international bankers. Additionally, French merchants from Le Havre, recruited by Madame Dufresne due to her prior connections with them, assisted in transporting the collection.[65] The operation also relied on numerous marginalized actors, including sailors, family members, and women who helped unpack the collection in Scotland. Overall, the sale and purchase of the collection involved 69 individuals with 142 connections between them.[66]

This network was not highly dense overall, with an average number of connections per actor of only four.[67] Two actors in particular, Dufresne and the naturalist Thomas Brown, used the laxity of the network to ensure successful transportation of the collection. Although Brown was merely an emissary from the University of Edinburgh sent to inspect the collection at the Parisian museum, he and Dufresne were the most important actors in the network.[68] Figures 4 and 5 illustrate the roles of Dufresne and Brown within the network. Dufresne's and Brown's importance resulted partially from the number of connections they had with other influential actors, such as Cuvier from the museum and Robert Jameson from the university.[69] Dufresne's and Brown's immediate networks were not densely connected, meaning the people they were connected to were

referenced indirectly in the letters, such as sailors, and women who presumably aided in unpacking the collection in Scotland. For the correspondence, see CRCEUL, Gen. 1801/2/Coll-198. When an unspecified number of people was mentioned (e.g., a ship crew, or "some ladies"), I added two nodes to the list.

64. On standardization in its international dimensions, see also Crosland, "Congress on Definitive Metric Standards"; Grossman, "Standardization (Standardisation)"; Kershaw, "International Electrical Units"; and Riordan, "Making of the Kilogram."

65. Brown to Jameson, Mar. 29, 1819, Gen. 1801/2/Coll-198, no. 16, CRCEUL. In terms of gender composition, the list of actors in the network is overwhelmingly male. Beyond Sarah Bowdich's (Lee) role in the network, the fact that she, as a woman, translated Dufresne's work is significant. On the gender aspect of her authorship, see Orr, "Stuff of Translation." On taxidermy and gender, see Haraway, "Teddy Bear Patriarchy."

66. To visualize the network, I used Gephi, an open-source software package for network analysis and visualization. On network analysis, see Knoke and Yang, *Social Network Analysis*.

67. This claim is also supported by network density, calculated by dividing the number of actual connections by the number of possible connections. If every node in the network is connected to every other node, the network density would be 1. Dufresne's network scores .061.

68. As Franco Moretti has demonstrated in the case of novels' narratives ("Network Theory, Plot Analysis"), networks can be helpful in revealing underlying structures, and in doing so they can help in challenging preconceived notions concerning hierarchies between and centrality of characters.

69. The PageRank measure evaluates the importance of each actor in a network by considering the number of incoming links and the significance of the actors that provide these links.

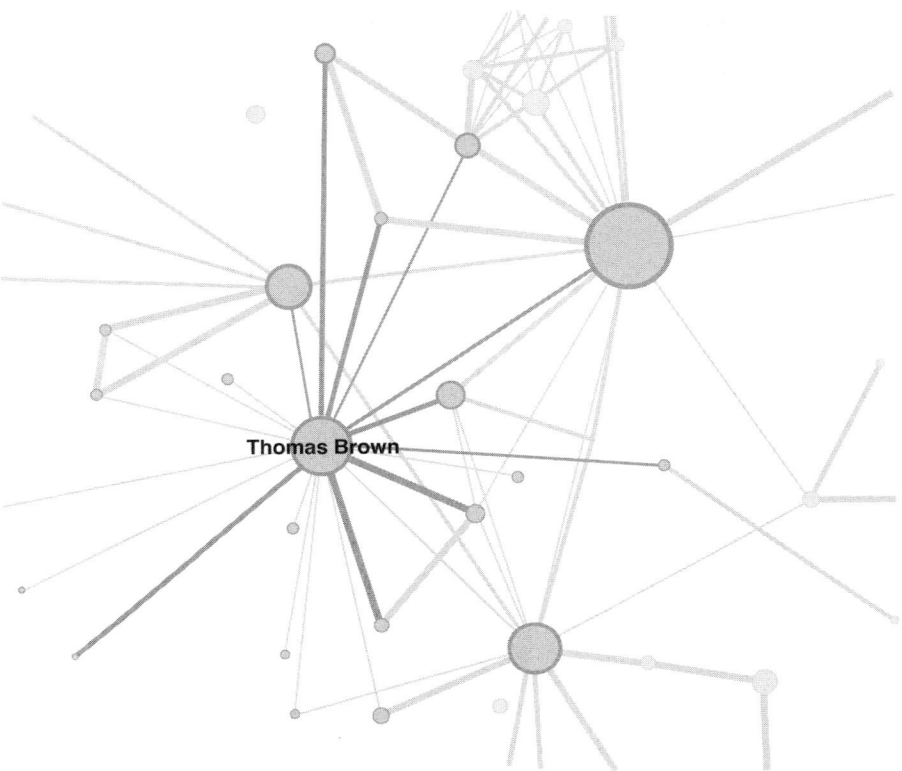

Thomas Brown

FIGURE 4 Brown's mastery of the network through loose ties. Bold lines represent direct connections to other actors; faded lines indicate connections between other actors. Node size reflects its number of connections.

not necessarily connected with each other.[70] Many of these connections are to service providers, such as a French insurance company. Working within the Parisian museum allowed Dufresne and Brown to develop a close relationship essential for the transaction while permitting each to maintain a loose web of connections with service providers, government institutions, and other entities in the city and its vicinity and in Britain.[71] Years later Brown described his close relationship

70. The clustering coefficient measures the density of immediate connections an actor has to the network. Dufresne scores 0.09; Brown, 0.06. Complete connection density would have a clustering coefficient of 1.

71. Dufresne and Brown also score high (first and third place, resp.) in betweenness centrality, which measures the shortest paths between nodes in the network, denoting how much information passes through that node. A higher betweenness centrality score means more information passes through that node, which denotes a greater degree of control over the network. For a study that uses network analysis to unearth actors beyond renowned philosophers, see Comsa et al., "French Enlightenment Network." See also Langmead et al. "Towards Interoperable Network Ontologies."

FIGURE 5 Dufresne's mastery of the network through loose ties. Lines and nodes are as in figure 4.

with Dufresne: "The intimacy that I then formed with M. Dufresne, one of the best naturalists in Europe, and situated as he was at the head of the laboratory of preservation, afforded me ample opportunity of inspecting and becoming acquainted with all the different processes employed in the preservation of animals."[72]

The diverse array of practices and exchanges taking place at the National Museum of Natural History contributed to the flexibility and dynamism of Dufresne's network. For example, William Macdonald, a professor of natural history at Edinburgh, was at the museum purchasing his own collection of minerals, shells, and a herbarium from one of the institution's botanists. Macdonald

72. Brown, *Taxidermist's Manual*, vi.

eventually collaborated with Brown and Dufresne, transporting his new collection on the same ship that carried Dufresne's specimens.[73] Macdonald's serendipitous presence at the museum allowed the Scottish university staff flexibility in selecting parts of the collection that suited their needs: Dufresne's collection included eleven books, which he offered for 4,000 francs. Macdonald chose to acquire these books for himself, intending to introduce them to Edinburgh's intellectual community. The university reserved the option to purchase the books from him at a later date for the same price.[74] Additionally, the museum had at least one staff member, A. Royer, proficient in English who assisted in translating relevant documents, handling logistical issues such as import taxes and customs clearance, and preparing a catalog of the collection. This catalog, elegantly bound in red Morocco, included empty pages after each genre listing for the university staff to add future information on the specimens.[75] Thus Royer, like Macdonald, played a crucial role in accommodating the needs of the emerging Scottish national museum. Higher-ranking officials at the Parisian museum also acted as facilitators. For instance, Cuvier added several quadruped specimens from the museum's official collection to the exported Dufresne collection as a gift to the University of Edinburgh.[76] These actions, enabled by the loose transnational network surrounding the museum, helped ease the complex process of integrating a private French collection into the national Scottish museum, at a time when public national museums were coming into being and had no transnational standards.

The acclimation of the collection to the aspirations of the Edinburgh museum took shape through a diverse array of media. The physical presence at the museum, including oral conversations and observation, was a crucial medium. Textual correspondence between Paris and Edinburgh was another. Material culture also played a role in this process of adjustment, exemplified by Cuvier's gift of quadrupeds. Dufresne contributed intertextual aids involving written and visual texts to all of these media. Although no transnational standards for the preservation and display of specimens existed at the time, the university acquired Dufresne's "classical" collection partly because of its association with the Parisian National Museum of Natural History, the first museum of its kind. Dufresne sought to ensure that the collection would maintain the standards of this museum. He provided the Scottish team with painstaking and detailed instructions for unpacking and redisplaying the collection. To ensure accurate reproduction in Edinburgh, he

73. William Macdonald to Robert Jameson, May 5, 1819, Gen. 1801/2/Coll-198, no. 25, CRCEUL; and Macdonald to Jameson, May 25 1819, no. 25a.

74. Macdonald to Jameson, May 25, 1819.

75. See Brown to Jameson, Mar. 29, 1819.

76. See Louis Dufresne to Robert Jameson, July 9, 1819, Gen. 1801/2/Coll-198, no. 58, CRCEUL.

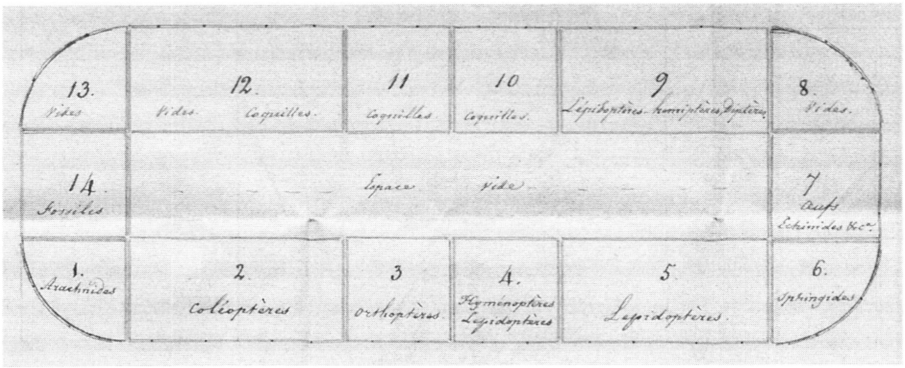

FIGURE 6 Dufresne's original arrangement of his collection (July 20, 1819, Gen. 1801/2/ Coll-198, no. 62, Centre for Research Collections, Edinburgh University Library). The University of Edinburgh Heritage Collections.

offered a plan detailing the original arrangement of the Parisian collection (see fig. 6).[77] In his instructions he stressed, for example, that when small birds were placed on shelves, they should be displayed "at a proper distance from each other, an inch or less, so that they don't touch." The fourteen chests of drawers containing the insects, he added, should be placed "in the middle of the room . . . or all around against the walls," with chest numbers 1, 6, 8, and 13 situated at the corners of the room.[78] His unpacking instructions were equally meticulous: he explained how to reconstruct a preserved specimen in the Parisian fashion. For example, Dufresne noted that the smaller birds, each placed in a separate roll of paper with the wire holding the legs bent back to save paper, should be taken out of the paper "from its smaller end," having the bird "redressed upon the stand by putting the thumb upon the feet of the bird and with the index placed behind its legs bringing the bird forward until it has reached its natural position." The insects in Dufresne's collection required even more painstaking care: the pins holding the specimens were to "be drawn off in the bent they have and with a proper instrument for it with the immediate side of the index and thumb."[79]

Based on the available sources, it appears that the university's team successfully maintained the cohesiveness of the Parisian collection. In 1825 the walls of the upper great room of the new Scottish museum, illuminated by three large

77. In Dufresne's plan (fig. 6), sections 10, 11, and 12 are devoted to shells. Shells, and invertebrates more broadly, played a crucial role in Lamarck's theory of biological change. As mentioned above, the shells in Dufresne's collection were arranged and named according to Lamarck's system. On Lamarck, see the seminal work of Burkhardt, *Spirit of System*. For more recent work, see Burkhardt, "Lamarck, Evolution"; and Gissis, *Lamarckism*.
78. See Louis Dufresne to Robert Jameson, July 20, 1819, Gen. 1801/2/Coll-198, no. 62, CRCEUL.
79. See Dufresne to Jameson, July 20, 1819.

lanterns from the roof and three grand windows from the side, were adorned with cases covered with plate glass displaying Dufresne's bird specimens. The cases beneath the gallery contained the university's existing bird collection. At the center of the room, which had an iron floor painted for aesthetic purposes, tables covered with plate glass showcased collections of shells, insects, and corals, most probably Dufresne's. The lower external gallery, a fifty-foot-long room, housed an extensive insect collection and a mineralogy cabinet for student use.[80]

The fluid transnational community surrounding the French National Museum of Natural History played an important role in enabling the movement across national borders of massive collections like Dufresne's. In this sense, this community constituted an extension of the French military and political might that facilitated the transportation of foreign collections such as the Dutch and Portuguese to Paris as war bounty. This place in the transnational scientific arena strengthened Paris's comparative advantage in the market for scientific specimens.

A Parisian Brand

Beyond the material advantages that Paris possessed—a large volume and wide variety of high-quality specimens and status as a hub of transnational scientific exchange—the French capital benefited from having high levels of symbolic capital. Paris's comparative advantage in the transnational market for specimens was reflected in foreigners acknowledging its tangible assets. The national museum in Paris was a brand in the making, and with it Dufresne was creating a transnational name for himself as *the* European authority on preservation.

The many foreign visitors that the museum attracted in those years were crucial for boosting the institution's transnational reputation. One English observer attributed the "splendid advances which have been made in our knowledge of the works of nature" to the Parisian museum. The author contrasted this to the lack of national funding and mismanagement in England, asserting that "our institutions for the encouragement of natural history" were in a "state of backwardness."[81] Another commentator remarked that "comparing Paris and London, with respect to wealth and population, the taste for natural history is much greater among the instructed classes, and more generally diffused among tradesmen and shopkeepers, in the former city, than in the latter." The examples he gave were the proliferation in Paris of "showy and odoriferous flowers" and "singing birds." For this commentator, the museum in Paris stood at the

80. See Stark, *Picture of Edinburgh*, 181–83.
81. *London Magazine*, 401–2.

forefront of this broad cultural engagement with nature. "The menagerie in the Jardin des Plantes, and the rich and extensive collections in the museum of the same establishment, it may easily be conceived," the observer continued, "have contributed to the taste for, and knowledge of, zoology."[82] He believed that taxidermy was a major French achievement: "It was observed to us, by an eminent naturalist here, that the French were greatly in advance of the English, in all that related to the preservation of the subjects of natural history; a fact which we think cannot be denied, on comparing by memory the preserved animals and birds in the British Museum . . . with the national collections here."[83] To use Richard W. Burkhardt's terminology, Britons perceived the French as experts in "civilizing" specimens, in making them "suited to the needs of science and society."[84]

More broadly, British newspapers, journals, and books portrayed the French as skillful creators of a natural world tailored to polite society. However, unlike the realm of science, which was perceived as requiring a certain degree of human-imposed order, in other domains these portrayals often criticized the artificial French approach to nature. What is important here is the portrayal of the French as promoters of a "civilized" version of nature. In reference to French gardening and genre painting, a Scottish observer opined that the French "never lose sight of the effort of the artist; their admiration is fixed not on the quality of the object in nature, but on the artificial representation of it; not on the thing signified, but the sign." The writer further commented that, "leaving the charming heights of Belleville, or the sequestered banks of the Seine, almost wholly deserted, they crowd to the stiff alleys of the Elysian Fields, or the artificial beauties of gardens of Versailles." While in Paris, this "artificial style of gardening" was "in unison, in some measure, with the regular character of the buildings with which it is surrounded," outside of the city "it destroys altogether the effect which arises from the irregularity of natural beauty."[85] Another observer offered a trenchant critique of Paris's obsession with refining nature by drawing an analogy between French gardening and theater, emphasizing that "their gardens, for instance, are like their tragedies, formal and artificial; both curb'd and spoil'd by rules not to be found in nature."[86] A French observer, whose letter appeared in a British periodical, similarly assessed how the French obsessively processed nature: "An Englishman seems content to love Nature herself, A Frenchman can

82. Loudon, *Magazine of Natural History*, 385.
83. Loudon, *Magazine of Natural History*, 386.
84. Burkhardt, "Keynote," 13.
85. Alison, *Travels in France*, 78–79. On the relationship between French and English gardening, see Weltman-Aron, *On Other Grounds*.
86. Cumberland, *British Drama*, xix–xx.

love Nature too; but his admiration of her increases in proportion as she calls up feelings connected with *himself*: just as he loves his wife or his mistress best when she happens to have on a dress that *he* chose for her."[87] It is no surprise, then, that in the preservation of organisms for display in museums—the most obvious manifestation of nature within polite society—the French were perceived as almost natural experts. In the eyes of at least one Anglophone observer, taxidermy was a quintessential French art. As he put it, "The art of taxidermy has not attracted as much attention in this country as might have been expected, considering the facilities which we possess for obtaining specimens from almost every country (for into what corner of the globe has not British enterprise penetrated?), and it must be confessed that in this study the French naturalists have taken the lead of the English."[88]

It is in this sociocultural context, in which Parisian polite society and its museum were branded as experts in fitting nature into culture, that we should examine the dissemination of Dufresne's methods abroad through the translation of his manual.[89] The translator, Sarah Bowdich (who remained anonymous in the first editions), and her then husband, the explorer Edward Bowdich, spent four years at the Parisian museum. During their stay, the couple became part of the museum's cosmopolitan community of naturalists and frequented, for instance, the salon of Anne-Marie Duvaucel, Georges Cuvier's wife, and their daughters, Sophie and Clementine. Subsequently, the couple became enthusiastic mediators of French science to Anglophone audiences.[90]

Bowdich was proud of her Parisian connections. The translation of the manual was more than a literary endeavor.[91] It conveyed Bowdich's own hands-on experience learning from Dufresne and the naturalists at the museum. Bowdich added to the translation an observation she made at the museum. In a note, she conveyed that she had observed how "the Artists in the Zoological Laboratory in Paris, carefully bend or turn down the points of the various wires, after they have inserted them (as they easily straighten them again with the fingers, if requisite) lest by pricking their fingers, the arsenic might do them serious injury."[92]

87. M. de Saint Foix, "Letters on England," Oct. 2, 1817, *New Monthly Magazine, and Literary Journal*, 279.

88. S. R., "On Taxidermy," *Mechanics' Magazine, Museum, Register, Journal, and Gazette*, 227.

89. On the Scottish Enlightenment, see Israel, *Democratic Enlightenment*, chap. 9. On Scottish Enlightenment thinkers' views of the French Revolution, see Plassart, *Scottish Enlightenment*.

90. Orr, "Women Peers in the Scientific Realm," 41–42.

91. Dufresne's dictionary entry was an ambitious project in its comprehensiveness. After providing a lengthy historical exposition of the development of taxidermy and an elaboration on the main travelers-preservers of the period, Dufresne enumerated the tools used in preservation and recipes for composing preservatives. He also addressed the preservation of specific species. See Dufresne, "Taxidermie," 514.

92. Bowdich (Lee), *Taxidermy*, 14.

Years later, in the text's sixth edition, in which she revealed her authorship, Bowdich (now adopting the surname Lee) assured readers that not only had she read the best writings on preservation of specimens and consulted skilled preservers but also "verified all my instructions in the laboratories of the Museum in Paris."[93]

Although Bowdich's 1820 translation did not name Dufresne as its original author, the Parisian taxidermist, and more so the French museum, were described in the translation as authorities on taxidermy and natural history more generally. Toward the end of the text, Bowdich declared that she had merely "given what has appeared to us most essential to the collection and preparation of objects of zoology." Those "who desire more detailed instructions" should consult "the Taxidermy of M. Dufresne, Chief of the Zoology Laboratories of the Museum, Paris." In the section devoted specifically to travelers, the translator stressed that it was "drawn up by the Professors of the Jardin du Roi, in Paris."[94]

Like Bowdich, Thomas Brown promoted himself in Britain as a mediator of French knowledge about the scientific representation of nature. In his *Taxidermist's Manual* from 1833 he highlighted the knowledge he had acquired while stationed at the Parisian museum in 1819. In addition to methods adopted from British naturalists such as William Bullock and Charles Waterton, the book includes a section devoted to techniques of skinning birds as practiced at the Parisian museum.[95] Reviews of Bowdich's translation generally highlighted its Parisian connection. Although the advertisements for the book did not mention Bowdich's connections to Paris, reviewers emphasized the "Frenchness" of the text and of taxidermy more generally. One reviewer, for instance, opened the review by noting that "taxidermie is a term recently introduced into the French language." They also explicitly attributed the term's usage to Dufresne. "The most approved directions for packing, dressing, and mounting animals of different descriptions," they noted, "are those which were published by M. Dufresne, assistant-naturalist and director of the zoological laboratories in the Museum of Natural History in Paris, in the first edition of the New Dictionary of Natural History, and which have been reprinted in the second edition of the work." "The numerous specimens in the Parisian Museum adjusted by M. Dufresne himself or by some of his pupils," stated the reviewer, "sufficiently attest [to] the superiority of the method over those which have been proposed by other writers on the same subject." As for the written text, although "his code of instructions," which was indeed "in many respects excellent," was "still remote from perfection," Dufresne had managed to compress "so much practical information within a

---

93. *Taxidermy; or, The Art of Collecting, Preparing, and Mounting Objects of Natural History*, iii.
94. *Taxidermy; or, The Art of Collecting, Preparing, and Mounting Objects of Natural History*, 137, 119.
95. Brown, *Taxidermist's Manual*, 28.

very limited number of pages, and detailed it in such a clear and perspicuous manner, that we are pleased to find his communication rendered accessible to the English reader by the appearance of the present volume."[96]

Another observer stated that "the attention of the French naturalists was for many years directed to the subject, and many treatises on Taxidermy, as is now called, were written." Dufresne's "great experience at the Museum at Paris" made his observations "of consequence."[97] Indeed, Dufresne was an authority to be reckoned with in Britain. In a review of the British naturalist William Swainson's *Naturalist's Guide for Collecting and Preserving Subjects of Natural History and Botany*, the reviewer compared Swainson with Dufresne. "In various important points," they wrote, "Mr. Swainson had been anticipated, with more ability and copiousness, by M. Dufresne, in his Treatise on Taxidermy; yet he has introduced some new valuable remarks, particularly with regard to the collection and preservation of shells, which the French writer had despatched in a more superficial manner."[98]

Thus, the new art of taxidermy entered the Anglophone world in a French wrapping, which remained unadulterated in the new context. In 1805, two years after the publication of Dufresne's first entry on taxidermy, one observer noted that "M. Dufresne, overseer of zoological operations at the laboratory of the National Museum of Natural History, has communicated an excellent essay on Taxidermia, or the art of dressing, stuffing, and preserving the skins of animals." The writer added that "for the sake of professed collectors and keepers of cabinets, we should be glad to see so many useful directions translated into English."[99] Bowdich's translation, which further popularized Dufresne's term for preservation, was followed by numerous texts of different kinds that referred to preservation under the new catchword *taxidermy*. With a readership that encompassed travelers, practitioners, and amateurs, Bowdich had cultural influence that probably surpassed the scale of visitors to the Edinburgh museum.[100] After the publication of Bowdich's translation, the use of the term proliferated in texts published across the Anglophone world, from the United States to India.

For example, in the *Manual of the Practical Naturalist*, published in Boston in 1831, an anonymous writer set out to enumerate the phases of collecting specimens, ranging from hunting to transportation and preservation. The author

96. *Monthly Review, or Literary Journal Enlarged*, vol. 93, 103–5.

97. Stark, *Elements of Natural History*, 503.

98. *Monthly Review, or Literary Journal Enlarged*, vol. 100, 210.

99. Review of *Nouveau dictionnaire d'histoire naturelle*, July 1805, in *Edinburgh Review*, 414.

100. By the time Bowdich translated the text into English in 1820, British taxidermists had produced their own preservation manuals (see, e.g., Bullock, *Concise and Easy Method*). But Dufresne's manual remained valuable, as can be witnessed in its multiple editions in English.

referred in the preservation segments to Dufresne's term *taxidermy*, which the writer defined as "the art of stuffing the skins of animals; and by extending the term, to that of mounting and preserving them, and restoring them to the appearance of life."[101] As for what preservatives to use, the writer left no room for doubt of their penchant for Dufresne's methods: "The best preservative against the ravages of insects, is that furnished by the naturalist Bécoeur [the method published by Dufresne]; his arsenical soap is used with success in the Museum of Natural History in Paris, and by all operators, traders, and amateurs in these articles in the capital."[102]

In 1830 the journal *Gleanings in Science* published a letter to the editor titled "On the Art of Taxidermy" by a British amateur writing in Upper-Doab, India. The writer explained that "the hill provinces abounding with birds of rare, if not unknown species, and some with the most beautiful plumage, has created a taste in travelers for collecting a set of specimens: and in order to encourage this pursuit, I have thrown together the following memoranda on the art of Taxidermy, which are at your service for publication." This writer also used Jean-Baptiste Bécoeur's method published by Dufresne.[103]

The newly coined term *taxidermy* and the methods Dufresne promoted enjoyed the association with the Parisian museum and with Paris more broadly. The perception of the French as the ultimate brokers between nature and polite society, and the museum's brand provided a convenient platform for the exportation of French methods abroad.

Conclusion

This article has adopted a transnational lens for examining the preserved dead body. Dufresne sold his collection at a time when the practices of preserving and collecting specimens were both highly national and cosmopolitan. A private collection associated with a national museum was particularly well suited to this environment of national-cosmopolitan science. As a private collection, it could be sold and transported with minimal intervention from the expanding French state bureaucracy. However, as part of the National Museum of Natural History's orbit, it benefited from the reputation and resources of France's well-funded scientific establishment, particularly given the museum's postrevolutionary image of producing universal scientific knowledge.

If the national museum can be likened to a central Broadway theater and Dufresne's collection to an Off Broadway show created by one of the Broadway

---

101. *Manual of the Practical Naturalist*, 95.
102. *Manual of the Practical Naturalist*, 85.
103. "On the Art of Taxidermy," *Gleanings in Science*, 160.

theater's employees, then the preserved animal body is akin to a character in a play. Much like the characters on a Broadway show, the preserved specimen is simultaneously real and fictional. It is real in the sense that it is embodied in an actual organism, but it is also an invention of the taxidermist's mind. Manufactured at the national museum, Dufresne's preserved specimens were acknowledged as being among the best in Europe, akin to talented actors on Broadway shows. Combining the agility of a private collection with the resources and reputation of a public museum, Dufresne's collection was a product of the specific circumstances of postrevolutionary France. As such, it serves as an apt example of France's competitive advantage in the market for natural historical specimens.

YOTAM A. TSAL is a fellow of the Dan David Society of Fellows at Tel Aviv University. He is author of "Newsworthy Nature: Capturing Exotic Birds in Buffon's France," *Eighteenth-Century Studies* 57, no. 3 (2024): 353–80.

Acknowledgments

The author wishes to thank Carla Hesse and Moshe Sluhovsky for their support and advice. He also wishes to thank the anonymous reviewers; the journal's editors, Christine Haynes and Jennifer Ngaire Heuer; and the editors of this special issue, April G. Shelford and Peter S. Soppelsa, for their thoughtful comments. The author also thanks Andy S. Chang, Snait B. Gissis, Zvi Hasnes-Beninson, Shaul Katzir, Josiane Lismay, and Shira Shmuely. He thanks for their generous assistance the staff at the Centre for Research Collections at the University of Edinburgh Library, and the staff at the National Museums Scotland Collection Centre in Edinburgh, particularly Vladimir Blagoderov, Milo Phillips, and Ashleigh Whiffin.

References

Alberti, Samuel, ed. *The Afterlives of Animals: A Museum Menagerie*. University of Virginia Press, 2011.

Alder, Ken. *The Measure of All Things: The Seven-Year Odyssey and Hidden Error That Transformed the World*. Free Press, 2001.

Alder, Ken. "Scientific Conventions: International Assemblies and Technical Standards from the Republic of Letters to Global Science." In *Nature Engaged: Science in Practice from the Renaissance to the Present*, edited by D. Kevles, M. Biagioli, and J. Riskin, 19–39. Palgrave Macmillan, 2012.

Alison, Archibald. *Travels in France, During the Years 1814–1815: Comprising a Residence at Paris During the Stay of the Allied Armies, and at Aix, at the Period of the Landing of Bonaparte*, vol. 1. 2nd ed. Edinburgh, 1816.

Aloi, Giovanni. *Speculative Taxidermy: Natural History, Animal Surfaces, and Art in the Anthropocene*. Columbia University Press, 2018.

Annales du Muséum d'histoire naturelle. Vol. 1. Paris, 1802.

Bell, David. *The First Total War: Napoleon's Europe and the Birth of Warfare as We Know It*. Houghton Mifflin Harcourt, 2007.

Bourguet, Marie-Noëlle, Christian Licoppe, and Heinz Otto Sibum, eds. *Instruments, Travel, and Science*. Routledge, 2002.

Bowdich (Lee), Sarah. *Taxidermy; or, The Art of Collecting, Preparing, and Mounting Objects of Natural History*. 6th ed. London, 1843.

Brown, Thomas. *The Taxidermist's Manual; or, The Art of Collecting, Preparing, and Preserving Objects of Natural History*. Glasgow, 1833.

Bullock, William. *A Concise and Easy Method of Preserving Objects of Natural History*. London, 1817.

Burkhardt, Richard W., Jr. "Lamarck, Evolution, and the Inheritance of Acquired Characters." *Genetics* 194, no. 4 (2013): 793–805.

Burkhardt, Richard W., Jr. *The Spirit of System: Lamarck and Evolutionary Biology*. Harvard University Press, 1977.

Burkhardt, Richard W., Jr. "Symposium Keynote Address: Civilizing Specimens and Citizens at the Muséum d'Histoire Naturelle, 1793–1838." *Of Elephants and Roses: French Natural History, 1790–1830*, edited by Sue Ann Prince, Memoirs of the American Philosophical Society, 267. APS Museum and American Philosophical Society, 2013: 12–28.

Chambers, Robert. *A Biographical Dictionary of Eminent Scotsmen*. Vol. 2. Glasgow, 1870.

Chappey, Jean-Luc. "Enjeux sociaux et politiques de la 'vulgarisation scientifique' en Révolution (1780–1810)." *Annales historiques de la Révolution française*, no. 338 (2004): 11–51.

Comsa, Maria, Melanie Conroy, Dan Edelstein, Chloe Summers Edmondson, and Claude Willan. "The French Enlightenment Network." *Journal of Modern History* 88, no. 3 (2016): 495–534.

Crawford, Elisabeth. "The Universe of International Science, 1880–1939." In *Solomon's House Revisited: The Organization and Institutionalization of Science*, edited by Tore Frängsmyr, 251–69. Science History Publications, 1990.

Crawford, Elisabeth, Terry Shinn, and Sverker Sörlin, eds. *Denationalizing Science: The Contexts of International Scientific Practice*. Kluwer, 1993.

Crosland, Maurice. "The Congress on Definitive Metric Standards, 1798–1799: The First International Scientific Conference?" *Isis* 60, no. 2 (1969): 226–31.

Crosland, Maurice. "A Science Empire in Napoleonic France." *History of Science* 44, no. 1 (2006): 29–48.

Cumberland, Richard. *The British Drama: A Collection of the Most Esteemed Dramatic Productions*, vol. 7. London, 1817.

Daston, Lorraine. "Nationalism and Scientific Neutrality Under Napoleon." In *Solomon's House Revisited: The Organization and Institutionalization of Science*, edited by Tore Frängsmyr, 99–115. Science History Publications, 1990.

Daston, Lorraine. "Type Specimens and Scientific Memory." *Critical Inquiry* 31, no. 1 (2004): 153–82.

Daudin, François-Marie. *Traité élémentaire et complet d'ornithologie*. Paris, 1800.

Delbourgo, James. *Collecting the World: Hans Sloane and the Origins of the British Museum*. Harvard University Press, 2017.

Deleuze, Joseph Philippe François. *Histoire et description du Muséum royal d'histoire naturelle*. Vol. 2. Paris, 1823.

Dufresne, Louis. *Catalogue des collections d'histoire naturelle formant le cabinet de M. L. Dufresne, chef du laboratoire de zoologie et naturaliste*. Paris, 1818.

Dufresne, Louis. "Taxidermie." In *Nouveau dictionnaire d'histoire naturelle appliquée aux arts principalement, à l'agriculture, à l'économie rurale et domestique*, 507–65. Paris, 1819.

Dufresne, Louis. "Taxidermie." In *Nouveau dictionnaire d'histoire naturelle appliquée aux arts principalement, à l'agriculture, à l'économie rurale et domestique, à la médecine*, vol. 32, 522–92. Paris, 1819.

Easterby-Smith, Sarah. *Cultivating Commerce: Cultures of Botany in Britain and France, 1760–1815.* Cambridge University Press, 2017.

Edinburgh Review. Vol. 6. Edinburgh, 1814.

Farber, Paul L. "The Development of Taxidermy and the History of Ornithology." *Isis* 68, no. 4 (1977): 550–66.

Farber, Paul L. "The Type-Concept in Zoology During the First Half of the Nineteenth Century." *Journal of the History of Biology* (1976): 93–119.

Findlen, Paula, ed. *Early Modern Things: Objects and Their Histories, 1500–1800.* Routledge, 2021.

Findlen, Paula. *Possessing Nature: Museums, Collecting, and Scientific Culture in Early Modern Italy.* University of California Press, 1994.

Fox, Robert. *The Savant and the State: Science and Cultural Politics in Nineteenth-Century France.* Johns Hopkins University Press, 2012.

Gascoigne, John. *Science and the State.* Cambridge University Press, 2019.

Gérardin, Sébastien. *Tableau élémentaire d'ornithologie, ou Histoire naturelle des oiseaux que l'on rencontre communément en France: Suivi d'un traité sur la manière de conserver leurs dépouilles.* Atlas, 1806.

Gillispie, Charles. *Science and Polity in France.* Princeton University Press, 2004.

Gissis, Snait B. *Lamarckism and the Emergence of "Scientific" Social Sciences in Nineteenth-Century Britain and France.* Springer Nature, 2024.

Gleanings in Science. Vol. 2. Edinburgh, 1830.

Goff, Alice. *The God Behind the Marble: The Fate of Art in the German Aesthetic State.* University of Chicago Press, 2024.

Grossman, Jonathan. "Standardization (Standardisation)." *Critical Inquiry* 44, no. 3 (2018): 447–78.

Habermeyer, Ryan. "Skin Walking." *Massachusetts Review* 63, no. 2 (2022): 317–27.

Haraway, Donna. "Teddy Bear Patriarchy: Taxidermy in the Garden of Eden, New York City, 1908–1936." *Social Text*, no. 11 (1984–85): 20–64.

Heilbron, John. *The Dilemmas of an Upright Man: Max Planck as Spokesman for German Science.* University of California Press, 1987.

Hendriksen, Marieke. "Animal Bodies Between Wonder and Natural History: Taxidermy in the Cabinet and Menagerie of Stadholder Willem V (1748–1806)." In Ross, "Preserving the Animal Body," 1110–31.

Israel, Jonathan. *Democratic Enlightenment: Philosophy, Revolution, and Human Rights, 1750–1790.* Oxford University Press, 2013.

Jansen, Justin. "The Bird Collection of the Muséum National d'Histoire Naturelle. Paris, France: The First Years (1793–1825)." *Journal of the National Museum* (Prague), Natural History Series, no. 184 (2015): 81–112.

Jansen, Justin. "Towards the Resolution of Long-Standing Issues Regarding the Birds Collected During the Baudin Expedition to Australia and Timor (1800–1804): A Review of Original Documents Reveals New Details About Collectors, Donors, Numbers, and Disbursement." *Journal of the National Museum* (Prague), Natural History Series, no. 183 (2014): 5–18.

Jones, Karen. "The Rhinoceros and the Chatham Railway: Taxidermy and the Production of Animal Presence in the 'Great Indoors.'" *History*, no. 348 (2016): 710–35.

Kershaw, Michael. "The International Electrical Units: A Failure in Standardisation?" *Studies in History and Philosophy of Science Part A* 38, no. 1 (2007): 108–31.

Knoke, David, and Song Yang. *Social Network Analysis.* Sage, 2019.

Lacour, Pierre-Yves. *La république naturaliste: Collections d'histoire naturelle et Révolution française, 1789–1804*. Publications scientifiques du Muséum, 2019.

Laissus, Yves. *Le Muséum national d'histoire naturelle*. Gallimard, 1995.

Langmead, Alison, Jessica M. Otis, Christopher N. Warren, Scott B. Weingart, and Lisa D. Zilinksi. "Towards Interoperable Network Ontologies for the Digital Humanities." *International Journal of Humanities and Arts Computing* 10, no. 1 (2016): 22–35.

Lipkowitz, Elise. "Seized Natural-History Collections and the Redefinition of Scientific Cosmopolitanism in the Era of the French Revolution." *British Journal for the History of Science* 47, no. 1 (2014): 15–41.

London Magazine. Vol. 6. London, 1826.

Loudon, J. C. *Magazine of Natural History And Journal of Zoology, Botany, Mineralogy, Geology, and Meteorology*. London, 1829.

Manual of the Practical Naturalist; or, Directions for Collecting, Preparing, and Preserving Subjects of Natural History. Boston, 1831.

Marte, Fernando, Amandine Péquignot, and David W. Von Endt. "Arsenic in Taxidermy Collections: History, Detection, and Management." *Collection Forum* 21, nos. 1–2 (2006): 143–50.

Mechanics' Magazine, Museum, Register, Journal, and Gazette. Vol. 19. London, 1833.

Miller, George. *Popular Philosophy; or, The Book of Nature Laid Open upon Christian Principles*. Dunbar, 1826.

Monthly Review, or Literary Journal Enlarged: From January to April, Inclusive. Vol. 100. London, 1823.

Monthly Review, or Literary Journal Enlarged: From September to December, Inclusive. Vol. 93. London, 1820.

Moretti, Franco. "Network Theory, Plot Analysis." *New Left Review*, no. 68 (2011): 81–86.

New Monthly Magazine, and Literary Journal. Vol. 3. London, 1822.

Nicolas, Pierre François. *Méthode de préparer et conserver les animaux*. Paris, 1800.

Orr, Mary. "The Stuff of Translation and Independent Female Scientific Authorship: The Case of *Taxidermy . . .*, Anon. (1820)." *Journal of Literature and Science* 8 (2015): 27–47.

Orr, Mary. "Women Peers in the Scientific Realm: Sarah Bowdich (Lee)'s Expert Collaborations with Georges Cuvier, 1825–1833." *Notes and Records: The Royal Society Journal of the History of Science* 69, no. 1 (2015): 37–51.

Osborne, Michael. "Applied Natural History and Utilitarian Ideals: Jacobin Science and the Muséum d'Histoire Naturelle." In *Re-Creating Authority in Revolutionary France*, edited by Bryant Ragan and Elizabeth Williams, 125–43. Rutgers University Press, 1992.

Outram, Dorinda. *Georges Cuvier: Vocation, Science, and Authority in Post-Revolutionary France*. Manchester University Press, 1984.

Parsons, Christopher M., and Kathleen S. Murphy. "Ecosystems Under Sail: Specimen Transport in the Eighteenth-Century French and British Atlantics." *Early American Studies* 10, no. 3 (2012): 503–39.

Patchett, Merle. "Putting Animals on Display: Geographies of Taxidermy Practice." PhD diss., University of Glasgow, 2010.

Péquignot, Amandine. "Histoire de la taxidermie en France (1729–1928)." PhD diss., Muséum National d'Histoire Naturelle, 2003.

Plassart, Anna. *The Scottish Enlightenment and the French Revolution*. Cambridge University Press, 2015.

Poliquin, Rachel. *The Breathless Zoo: Taxidermy and the Cultures of Longing*. Pennsylvania State University Press, 2012.

Riordan, Sally. "The Making of the Kilogram, 1789–1799." PhD diss., Stanford University, 2013.

Rookmaaker, L. C., P. A. Morris, I. E. Glenn, and P. J. Mundy. "The Ornithological Cabinet of Jean-Baptiste Bécoeur and the Secret of the Arsenical Soap." *Archives of Natural History* 33, no. 1 (2006): 146–58.

Ross, Alan S. "The Animal Body as Medium: Taxidermy and European Expansion, 1775–1865." *Past and Present*, no. 249 (2020): 85–119.

Ross, Alan S., ed. "Preserving the Animal Body: Cultures of Scholarship and Display, 1660–1914." Special issue, *Journal of Social History* 52, no. 4 (2019).

Scots Magazine and Edinburgh Literary Miscellany. Vol. 84. Edinburgh, 1819.

Spary, Emma. "On the Ironic Specimen of the Unicorn Horn in Enlightened Cabinets." In Ross, "Preserving the Animal Body," 1033–60.

Spary, Emma. *Utopia's Garden: French Natural History from Old Regime to Revolution.* University of Chicago Press, 2010.

Stark, John. *Elements of Natural History.* Vol. 1. Edinburgh, 1828.

Stark, John. *Picture of Edinburgh.* Edinburgh, 1825.

Swainson, William. *Taxidermy: Bibliography and Biography.* London, 1840.

Sweet, Jessie M. "The Collection of Louis Dufresne (1752–1832)." *Annals of Science* 26, no. 1 (1970): 33–71.

Sweet, Jessie M. "William Bullock's Collection and the University of Edinburgh, 1819." *Annals of Science* 26, no. 1 (1970): 23–32.

Taxidermy; or, The Art of Collecting, Preparing, and Mounting Objects of Natural History. London, 1820.

Tsal, Yotam. "Newsworthy Nature: Capturing Exotic Birds in Buffon's France." *Eighteenth-Century Studies* 57, no. 3 (2024): 353–80.

Van Damme, Stéphane. *Paris, capitale philosophique: De la Fronde à la Révolution.* Jacob, 2005.

Weltman-Aron, Brigitte. *On Other Grounds: Landscape Gardening and Nationalism in Eighteenth-Century England and France.* State University of New York Press, 2001.

Wonders, Karen. *Habitat Dioramas: Illusions of Wilderness in Museums of Natural History.* Uppsala University Press, 1993.

The Illumination of Restoration Paris

ELIZABETH DELLA ZAZZERA

ABSTRACT   Around 1816 French industrialists began debating gaslight, with concerns rang-ing from how the new technology would affect France's seed oil industry to the role that coal should play in plans for French industrialization. In 1823 the debate spilled into the public consciousness when the construction of a new gasometer provoked worries about the risk of explosion. By examining the gaslight controversy in the broader cultural context of the Restoration, and by exploring its intersections and parallels with other conflicts of the era, this article shows how illumination became about more than technology, visibility, or indus-trialization and highlights the fluidity of the meanings of gaslight. The debate went beyond a simple opposition of pro-technology liberals and antimodern reactionaries; gaslight took on many changing meanings throughout the conflict. Opposing gas could be prudent and ratio-nal, while supporting it could be unpatriotic and reckless. Gaslight could be seen as an aes-thetic affront or a spectacular innovation; it could be imagined as a threat to French sover-eignty or as a tool of the state.

KEYWORDS   gaslight, Romanticism, classicism, technology, industrialization

The Paris gaslight controversy began in 1823, when gas lamps were neither brand-new nor ubiquitous. The city's first public gaslights had been installed in 1816 in a shopping arcade, the Passage des Panoramas, but most public light-ing continued to be supplied by oil lanterns through the early 1840s.[1] Yet by 1823 gas lamps seem to have become just visible enough to spark fierce public debate. Gaslight, its detractors insisted, was smelly, ugly, dangerous, and foreign. For its proponents, gas seemed an obvious improvement on oil. The light it created was brighter, more controllable, more convenient, and more modern.

Parisians had fought over lighting before, once when Louis XIV first intro-duced street lighting in the form of candle lanterns in 1667 and then again when the Argand oil lamp was invented in the 1780s.[2] They would debate it again when the city began replacing the no-longer-controversial gas lamps with electric

---

1. Gas lamps made up 36.5 percent of streetlights in Paris in 1841 and 57.4 percent in 1843. By 1852 oil lanterns had been reduced to only 2.1 percent of all streetlights. Delattre, *Les douze heures noires*, 128.
2. McMahon, "Illuminating the Enlightenment."

French Historical Studies • Vol. 48, No. 2 (May 2025) • DOI 10.1215/00161071-11626745
Copyright 2025 by Society for French Historical Studies

lights in the late 1870s.[3] Each successive wave of new lighting technology conjured detractors, who insisted the old ways were best, and modernizers, who wanted to push forward with the innovation. Lighting (*éclairage*), with its obvious connection to light (*lumière*) and its accompanying metaphors (enlightenment), seemed especially poised to serve as a stand-in for larger conflicts about knowledge, reason, and modernity.[4]

But while the conflicts engendered by these moments of lighting innovation may rhyme, they did not repeat themselves. The 1823 gas lamp debate, not merely a generational conflict of scary new technology versus the safe good old days, was a Restoration debate, mired in the specifics of post-Napoleonic economics and politics. French industrialists began debating gaslight in earnest around 1816, and their concerns ranged from how the new technology would affect France's seed oil industry to the role coal should play in plans for French industrialization. In 1823 the debate spilled into the public consciousness. In Jean-Baptiste Fressoz's excellent account of the controversy, he attributes the popularization of the debate to a political scandal involving the contract to build a gasholder and to growing concerns about the risk of explosion.[5] Fressoz's account does mention the important cultural significances tied up with gaslight, but they are not its focus. This article takes up the question of how this controversy fit into broader cultural conflicts of the Restoration and how gaslight transformed Parisian nightlife, especially the theater. In doing so, it examines the instability of the meanings of gaslight. The gaslight debate reflected no simple alignment of pro-technology liberals and antimodern reactionaries—each side of the gas lamp debate could be dressed up to suit the aims of its partisans. Opposing gas could be prudent and rational, while supporting it could be unpatriotic and reckless. Gaslight could be seen as an aesthetic affront or a tool for spectacular innovation; it could be imagined as a threat to French sovereignty, or as a tool of the state.

In the Restoration, illumination became about more than technology, visibility, or industrialization. It became part of the cultural and political project of Restoration France, which was inextricably bound with the question of what *restoration* meant, of what it meant to *restore* not just the monarchy but France itself.[6] In this moment, every debate about literature, politics, and technology became wrapped up in a larger debate about the future of France and about its

3. Clayson, *Illuminated Paris.*
4. I follow Hollis Clayson in making the distinction between *éclairage* and *lumière* (*Illuminated Paris*, 3).
5. Fressoz, "Gas Lighting Controversy," 731–32.
6. See Boutry, "Restauration."

past. What was the best way forward for a country as it emerged from a period of twenty-five years of revolution and dictatorship? How much should it (or could it) try to look like the country that came before that period of change and disruption? What did it mean to be modern? Lighting was one terrain on which the French contested these questions.

Another terrain was literature. The primary literary conflict of the Restoration, the *bataille romantique*, is most succinctly described as a conflict between the new literature (Romanticism) that eschewed universal rules in favor of particularism and the old literature (classicism) that embraced the three unities and represented the height of French literary glory.[7] But this new/old characterization belies the fact that neither genre had a monopoly on political modernity. While some conservative royalists embraced classicism as the literature of the ancien régime, some liberals considered it the literature of Enlightenment and reason. While some liberals supported Romanticism because it offered freedom from the strictures of tradition, both political and literary, some royalists too saw Romanticism as liberating. They simply believed that the monarchy represented the true and stable path to liberty, rather than the false liberty of Jacobinism. To further complicate matters, these groups were not static, especially not in the early 1820s, when the French Romantic movement was still nascent. In 1823 Victor Hugo, for example, belonged both to the *cénacle* of royalist Romantics who published the journal *La muse française* and to the staunchly royalist (and predominantly classicist) literary organization the Société des Bonnes-Lettres. By 1830 Hugo had not only fully publicly embraced Romanticism but also definitively identified it with liberalism.[8] Yet, despite its limitations in describing the messiness of reality, the new/old, revolution/counterrevolution formulation permeated many contemporary discussions, which meant that debates about things that seemed distinct from each other, such as literature and illumination, could become entangled.

On October 2, 1823, the satirical literary daily *Diable boiteux* published a news spot about the Théâtre du Gymnase. The night before, due to a problem with the newly installed gas lamps, the theatergoers were "plunged into profound darkness." In recounting this incident, the satirical daily took a jab at a rival literary organization. When the lamps went off, they wrote, theatergoers "could have thought themselves transported to the rooms of the Société des Bonnes-Lettres."[9] At seven o'clock the problem (a shortage of gas, because of a

---

7. See Della Zazzera, "Romanticism in Print."

8. In the preface to his 1830 play *Hernani*, Hugo wrote that Romanticism was "*liberalism* in literature" (ii).

9. *Diable boiteux*, Oct. 2, 1823.

late delivery) was fixed and, the paper reported, the gaslight was turned on "brighter than ever" and "cast torrents of light on its dark blasphemers."[10] This is not the only example of the writers at the *Diable boiteux* making fun of the Société des Bonnes-Lettres ostensibly for holding their meetings in a room that was not well lit, loving the darkness, or being literally "unenlightened." The next month the *Diable boiteux* made fun of the society's supposed response to Louis Daguerre and Charles Bouton's diorama of Holyrood chapel, a large-scale painting of the ruins of the chapel lit from behind in such a way that the scene appeared to change over several minutes, revealing a figure in the shadows. The *Diable boiteux* wrote that the society so loved the "terrible darkness" of the diorama that they wanted to hold their meeting in the ruined chapel, but "only asked that Daguerre darken a little more the moon's rays."[11] This comment implied that the society was gloomy and depressing, that it was behind the times, that it rejected new technology without cause, and perhaps even that it did not understand the point of a diorama: to explore the visual effect of variations in light intensity. Notably, it was new lighting technology like gas lamps, as well as advances in optical science, that made the visual effects of the diorama possible.[12] The society, the *Diable boiteux* implied, would be happy to discuss royalism and classicism in the ruins of a chapel, as long as someone turned off the gas lamps.

Given the limited reach of gaslight infrastructure in 1823 Paris, it is very likely that the society did not have gaslights in their halls and instead relied on less bright lanterns. But the actual means by which they lit their meeting rooms or, indeed, how dark they were was not the point. By repeatedly referring to the society's dimly lit chambers and love of darkness, the *Diable boiteux*, which expressed political positions much more liberal than the society's, intended to mock the society's character and not the physical darkness of its assembly rooms. It also seems possible that they wanted to link the society, with its royalist politics and its classicist literary leanings, to the anti-gaslight movement— were the society members gaslight's "dark blasphemers"? Perhaps they intended to ridicule the society for a purported antimodern opposition to gas lamps and to advertise the paper's contrasting "enlightened" position. The *Diable boiteux* had published in support of gaslight in the past. In September 1823 they wrote

---

10. *Diable boiteux*, Oct. 2, 1823. That final phrase is a play on the line from Jean-Jacques Lefranc de Pompignan's "Ode sur la mort de J.-B Rousseau": "Le dieu [the sun], poursuivant sa carrière, / Verse des torrents de lumière / Sur ses obscurs blasphémateurs." This deliberate misquote also appeared in the play *Le magasin de lumière*, in the Romantic authors Charles Nodier and Amédée Pichot's pamphlet on the evils of gaslight, and in another article in *Diable boiteux*.
11. *Diable boiteux*, Nov. 9, 1823.
12. Tresch, *Romantic Machine*, 141.

that "the use of hydrogen gas as a means of lighting will be one of the most glorious headlines of the nineteenth century."[13]

But was the society anti–gas lamp? It is difficult to say for certain. *Les annales de la littérature et des arts*, the organ of the Société des Bonnes-Lettres, had published a theater review that July criticizing a production of *Tancrède* for plunging the audience into total darkness during a prison scene. The reviewer called this technique, which he noted was made easy because of gaslight, "charlatanism" and insisted that he would be even more justified in criticizing the usage if it were to become common in all the theaters that adopted gas lighting.[14] This review criticized the use of gaslight onstage and implied that the theater embraced the lighting effect because they could, without properly considering whether they should. But an earlier issue of the journal made note of the introduction of gas lamps to a theater without comment.[15]

Regardless of the society's actual position on gas lamps, associating them with darkness and with the extinguishment of gas lamps clearly carried enough rhetorical weight that the writers of the *Diable boiteux* thought it was worth making the joke more than once. That decision and the jokes themselves become more understandable if we place them in the context of the public debate about gas lamps, a context that contemporary audiences would have been able to supply themselves, particularly by October 1823, when the gaslight controversy was at its height.[16] Viewing these jokes in that context forces us to consider how the gas lamp debate touched on questions of literature, culture, and politics, and reconstructing that context allows us to see the Restoration as a generative era of possibility and not a reactionary return to the past. To reconstruct that context, this article explores the cultural valences of light, the development and implementation of this new technology, the places where its debates intersected with or paralleled conflicts in literature and advancements in theater, and the political meanings it accrued during the 1820s.

. . .

The French words *lumière* and *obscurité* both reference physical and metaphorical light/darkness. The Enlightenment was called *les Lumières*, but the word also referred to art, intelligence, and advancement more generally, and in the nineteenth century it was especially associated with what we would now call scientific knowledge. Scholars and critics in the Restoration used *les lumières* as a kind of

---

13. *Diable boiteux*, "Du gaz hydrogène et du Faubourg-Poissonnière."
14. Le Vieil Amateur, "Spectacles," 34.
15. *Annales de la littérature et des arts*, "Théâtre Royal Italien," 190.
16. On humor as a historical lens, see Darnton, *Great Cat Massacre*.

catchall for knowledge-based labor, but it could also be evaluative. The *progrès des lumières* could mean the development of a particular vision of the right kind of scholarship, writing, and knowledge, the precise parameters of which varied depending on the interlocutor's aesthetic and/or political affiliation. During the Restoration, optics, the scientific study of light, saw innovations and debates that affected not only practical engineering concerns but also, as Theresa Levitt argues, conceptions of representation—of what an observer could actually see.[17] Light could encompass not only technological, intellectual, or political but also artistic advancement. Many of the new artistic media developed in the early nineteenth century—the panorama, the diorama, the polyorama panoptique, and the magic lantern—hinged on the application and control of light.[18] And the line between the science of illumination and its artistic application was not stark. For example, the eighteenth-century chemist Antoine-Laurent de Lavoisier, who made important contributions in the development of new lighting technologies, used his technical expertise to write a treatise on how to best light a theater.[19]

Paris is, of course, *la Ville-Lumière*, a phrase whose origin is generally traced to Paris's role as a center of the Enlightenment, although the term itself seems to have entered common parlance only in 1876.[20] But its meaning was always metaphorical and literal, even *avant la lettre*. As Darrin M. McMahon has shown, in the *siècle des Lumières* many philosophes insisted that intellectual enlightenment and physical illumination went hand in hand—"a lighted city was an enlightened city."[21] And after the Revolution, Simone Delattre tells us, Paris was (or was imagined to be) "the capital of the material and moral flourishing [*l'épanouissement*] of humanity, a beacon dispensing light destined to shine on all," a vocation that became incompatible with the dark streets of a bygone time.[22] Every push to expand street lighting, she argues, was a victory against resignation to darkness by a Paris that wished to demonstrate its reputation as the city of Enlightenment holding the "torch of civilization."[23] This mixing of metaphor and reality seems to accompany each attempt by French authorities to illuminate Paris streets and chase away the darkness.

---

17. Levitt, *Short Bright Flash*; Levitt, *Shadow of Enlightenment*.
18. Schivelbusch, *Disenchanted Night*, 213.
19. Lavoisier, "Mémoire sur la manière"; McMahon, "Illuminating the Enlightenment," 132.
20. The concept of Paris as *ville de lumière* seems to have been popularized by Hugo's collection of writing from his time in exile, published in 1875: "Whoever looks for progress sees Paris. There are black towns; Paris is *la ville de lumière*" (*Actes et paroles*, xlvi). The hyphenated version of the term may have originated in the newspaper press. The first instance I find is in a newspaper article from January 1876: a seemingly mocking reference to Hugo as the "délegué de la Ville-Lumière" (Loth, *Univers*).
21. McMahon, "Illuminating the Enlightenment," 136.
22. Delattre, *Les douze heures noires*, 115. *Epanoui* also means "glowing" or "radiant."
23. Delattre, *Les douze heures noires*, 115.

Paris's forays into municipal lighting predate the introduction of gas lamps by a century and a half and tell a story of technological change and expanding brightness. In 1667 Louis XIV issued a decree that candle lanterns be hung in the streets, which made Paris the first European city to introduce public street lighting. Eventually, thousands of candle lanterns illuminated the streets of Louis XIV's Paris, initially only in the winter months but later in the fall and spring as well. Wide-scale public lighting was intended to prevent crimes committed in the dark of night, such as robberies and murders, and was instituted the same year the state established a new police force for Paris. It should thus be seen as part the state's attempts to control and civilize a purportedly unruly city.[24] In the mid-eighteenth century the state replaced candles with oil lanterns, and a discourse emerged conflating brighter light and new lighting technologies with progress and erudition. A 1755 satirical essay on the history of lanterns replicated this rhetorical conflation, noting that we have given "the name *Lanterne par excellence* to sages who have brought light to some art or some science."[25] In 1766 the city introduced *réverbère*-style oil lamps (with mirrors to reflect the light) in the streets of Paris, each of which produced as much light as five or six regular oil lanterns, allowing the city to reduce the total number of lanterns while maintaining similar or even greater levels of luminosity.[26] The next significant advancement in oil lantern technology came in the 1780s with the Argand lamp. Designed by François Ami Argand, the lamp made use of Lavoisier's findings about the importance of air for combustion and featured a hollow wick that allowed air to enter the flame from both the inside and the outside. The Argand lamp burned brighter and more efficiently than earlier oil lamps.[27] Progress in illumination, in the vibrancy of light, remained a symbol of scientific and intellectual advancement. In 1781 Louis Sebastien de Mercier wrote in his *Tableau de Paris* that the twelve hundred *réverbères* in Paris offered consistent, durable clarity to the Parisian nights (except during full moons when the lanterns stayed dark),[28] and a few years later the allegorical text *Les veilles lanternes, conte nouveau* (1785) presented *réverbères* as "a symbol for the marvels of modern science."[29] Illumination continued to grow and spread across the city, and by 1817, when public gas lamps had just begun to appear in Paris, there were over forty-six hundred oil *réverbères* in the city.[30] While illumination continued to

24. McMahon, "Illuminating the Enlightenment," 124–26.

25. Dreux du Radier et al., *Essai historique, critique, philologique, politique*, 131; McMahon, "Illuminating the Enlightenment," 134.

26. Schivelbusch, *Disenchanted Night*, 95; McMahon, "Illuminating the Enlightenment," 129.

27. Schivelbusch, *Disenchanted Night*, 9, 11.

28. Mercier, *Tableau de Paris*, 116.

29. McMahon, "Illuminating the Enlightenment," 134.

30. Bogard, *End of Night*, 54.

be inequitably distributed across Paris neighborhoods through the 1850s,[31] in the Restoration, even before the introduction of gaslight, Parisians lived in a city of light.

As the many scholars of nocturnalization—"the ongoing expansion of the legitimate social and symbolic uses of the night"[32]—have demonstrated, public lighting changed people's relationship to the night,[33] even as people continued to complain about low visibility for nighttime travel.[34] Public lighting meant greater public safety and increased opportunities for social and commercial interaction after dark.[35] Evening mealtimes shifted later both at home and at cafes and coffeehouses.[36] In Restoration Paris, *cabinets de lecture* were open late into the evening, often until eleven o'clock, contributing to late-night sociability based around reading and the accumulation of knowledge. One travel guide described these reading rooms as places people went to spend the evening "tiring their eyes on the bright lights [*lumières*] that diffuse from the daily newspapers and the gas,"[37] continuing the tradition of using language to conflate the light of knowledge with the light of illumination.

Gas lighting marked a significant technological departure from oil lanterns. Gaslight illuminated streets and rooms as the final step of an industrial process that involved the manufacture, shipping, and burning of coal gas or wood gas. That process meant that the journey to produce gaslight began in coal mines and moved through factories, storage facilities, and pipes laid underground and inside walls before arriving at its destination inside gas lamps, waiting for a lamplighter. It also meant that the transition to gas lighting required a significant investment in infrastructure, including gasometers for storage, pipes for transportation, and new lampposts and light fixtures made to burn gas and connected to the gas supply pipes.[38] This new technology also transformed the sensory experience of artificial light. Gas burned hotter and therefore brighter than oil, and the flow of the gas and the strength of the light could be regulated, giving people some measure of control over brightness (although irregularities in the level of supply of gas could also affect brightness). While gas lamps have

---

31. Delattre, *Les douze heures noires*, 145.
32. Koslofsky, *Evening's Empire*, 2.
33. See, e.g., Schivelbusch, *Disenchanted Night*; Delattre, *Les douze heures noires*; Koslofsky, *Evening's Empire*; Schlör, *Nights in the Big City*; and Zallen, *American Lucifers*.
34. Guillerme, "Enclosing Nature in the City," 83.
35. McMahon, "Illuminating the Enlightenment," 128.
36. Bogard, *End of Night*, 54.
37. Pain and Beauregard, *Nouveaux tableaux de Paris*, 65–66.
38. On the significance of building a gas network and the technological infrastructure required, see Williot, "Naissance d'un reseau gazier à Paris au XIXème siècle"; and Tomory, "Building the First Gas Network."

luminescence similar to that of a forty-watt bulb,[39] contemporary descriptions of gaslight often note its incredible brightness—a French traveler to Birmingham remarked that a factory lit by gas lamps was lit "as though by the sun."[40] To someone used to candlelight or oil lanterns, gaslight would have seemed significantly brighter in comparison.

Unlike in England, where gas lighting was initially developed for industry,[41] in less-industrialized France gaslights first appeared in individual homes. In 1801 Philip Lebon, the inventor of the thermolamp, which he intended as a source of both light and heat, set up gas lighting in his home and charged admission for curious visitors.[42] However, it would take more than a decade for Paris to begin manufacturing gas for lighting on a large scale. Frederick Albert Winsor, inspired by tales of Lebon's thermolamp, started the Gas Light and Coke Company in London, which had begun lighting its streets with gas lamps in 1807.[43] In 1815 Winsor brought gas manufacture to Paris.[44] Although Winsor received an import license from the French state, his plans initially met with some public opposition and fears. In an attempt to win people over he placed gas lamps in the Passage des Panoramas, a shopping arcade, and in a salon that he opened to public viewing in spring of 1816. He later explained that he intended these public demonstrations as a visual rebuttal to French concerns about gas lamps (having learned in England the limitations of verbal rebuttals), suggesting that he believed that the simple experience of gas lamps would be enough to convince anyone of their superiority.[45] While that turned out not to be the case, Paris began replacing its street lighting with gas lamps in 1819—a process that would take decades.[46] Also in 1819, the Opéra became the first theater to install gas lighting.[47]

Even though gas lamps made up a minority of public lighting through the first half of the nineteenth century, as a new technology they sparked significant controversy. In Paris that controversy came to a head in 1823 over the enormous two-hundred-thousand-cubic-foot gasometer built by Antoine Pauwels's Compagnie Française de l'Eclairage par le Gaz in a very fashionable and wealthy Paris neighborhood on the Rue Faubourg-Poissonnière.[48] The practical concerns about the construction of such a large and potentially dangerous structure

39. Bogard, *End of Night*, 42.
40. Simond, *Voyage d'un français en Angleterre*, 127.
41. Schivelbusch, *Disenchanted Night*, 16.
42. Schivelbusch, *Disenchanted Night*, 23.
43. Fressoz, "Gas Lighting Controversy," 751.
44. Guillerme, "Enclosing Nature in the City," 83.
45. Winsor, *Résumé historique et démonstratif*, 10.
46. Allemagne, *Histoire du luminaire*, 526–27.
47. Allemagne, *Histoire du luminaire*, 523.
48. Fressoz, "Gas Lighting Controversy," 732.

intersected with partisan politics. The company had been granted permission to build two years earlier by the liberal government led by Elie Decazes, which had since fallen. Those who opposed the gasometer managed to convince the ultraroyalist Joseph de Villèle government to overturn the previous decision. The liberal press presented the Compagnie Française as a patriotic movement against British interests in the French gas lighting industry and claimed the ultraroyalists were playing politics because the Compagnie Française was financed by several elite liberals in France.[49] But the controversy was about changing ways of life as much as it was about political jockeying. As Fressoz demonstrates, unlike other industrial technologies, which existed far from fashionable neighborhoods, hidden away in factories, the inconveniences of gas lighting affected everyone. The more gas lamps appeared in public spaces—theaters, arcades, cafés, *cabinets de lecture*—the more Parisians encountered its potential downsides and dangers in their daily lives and the louder the debate grew.[50] As Fressoz makes clear, while the residents who opposed the gasometer in their wealthy neighborhood may have been nineteenth-century not-in-my-back-yarders, their fears and concerns were not without foundation: gas holders sometimes exploded, they could be large and unsightly, and gas was malodorous and potentially toxic.

We are primed to think of new technology as progressive simply because it is new, and of the people who oppose it as unnecessarily conservative or even reactionary. This maxim is especially true for old technologies that we no longer use and now seem innocuous. Why would anyone oppose gas lamps? To our sensibilities, gas lamps seem as nonthreatening as candles, but even more interesting for being rarer. However, opponents of gas lamps represented more than knee-jerk technological reactionism, and not just because the risks associated with gas lamps were real.[51] It is true that much of the public furor around gas lamps faded after the government passed a royal ordinance regulating them on August 20, 1824. The ordinance instituted a number of safety measures, designed to mitigate the possibility of a gas explosion[52] and placed gas factories under the surveillance of the police to ensure that they conformed to the regulations.[53] But the debate over gas lamps transcended worries over health and safety or technological risk. Like earlier lighting debates, it got wrapped up in questions of progress. Moreover, this technology that lighted spaces of reading and culture engendered

49. Fressoz, "Gas Lighting Controversy," 732.
50. Fressoz, "Gas Lighting Controversy," 732.
51. Fressoz, "Gas Lighting Controversy," 730.
52. Fressoz, "Gas Lighting Controversy," 743. Fressoz notes that one major outcome of the gas lamp debate was the implementation of these safety regulations. The gas industry was much more tightly regulated in France than in Britain, where the controversies had been less acute.
53. Any factory found in noncompliance was then shut down. Duvergier, *Collection complète des lois*, 581.

controversy that was itself cultural, and its partisans made their positions legible by discussing gaslight in tandem with their other concerns, including aesthetic ones.

In the minds of the pro-gaslight contingent, any opposition to this technological advancement was an opposition to progress and the advancement of truth. But they also frequently expressed certainty that the truth would prevail. In an article on the controversy, the *Diable boiteux* remarked that "the most useful discoveries always find adversaries ready to countermine its progress" and compared opponents of gaslight to cowpox inoculation (*la vaccine*) skeptics and enemies of mutual education (a pedagogic innovation of the eighteenth century that gained traction under the Restoration but saw opposition from the Catholic Church and from royalists). The article also insisted that disinterested investigations into the truth would vindicate gaslight against the accusations of its opponents. It then offered the same deliberate misquote of Lefranc de Pompignan's poem mentioned above—"the gas, pursuing its course, casts torrents of light on its dark blasphemers."[54] In the original poem, it is the sun, after having been insulted by a group of foolish barbarians, that casts its light on those unbelievers. With this parallel the author seemed to suggest that gaslight would persist despite its detractors. Even its enemies would be showered in its light, maybe even recognizing their error in the face of its brilliance. Similarly, Winsor's 1824 *Résumé historique et démonstratif sur l'éclairage par le gaz hydrogène* presented opposition to gas lighting as the result of prejudice, ignorance, and error. Gaslight, Winsor began, was a settled issue for scientists but continued to be a subject of concern for the public, whose fears he attributed to "ignorance of principles."[55] Winsor understood that gaslight, like any innovation in science or technology, would encounter obstacles, acknowledging that "we renounce old habits with difficulty; the aversion increases when those habits are tied to life's daily practices." But he believed that after the introduction of any innovation, truth and the common good (*l'intérêt général*) would always triumph, explaining that as "distrust gives way to the demonstrations of reason, to the even more powerful lessons of experience, chimerical fears vanish, the real advantages are better appreciated day by day." This fate, he insisted, would befall gas lighting in France, where it already had the support of experts, and would "gather all the votes in its favor," as anti–gas lamp prejudice yielded to the truth.[56] Winsor thus claimed not only that supporting gas lighting was more rational and more enlightened but also that the very rationality of the position—the power of its truth—ensured its eventual practical success. Winsor based his prognostication on prior experience in England, where, he explained, interested and prejudiced opposition to

54. *Diable boiteux*, "Du gaz hydrogène et du Faubourg-Poissonnière."
55. Winsor, *Résumé historique et démonstratif*, 1.
56. Winsor, *Résumé historique et démonstratif*, 30.

gaslight had given way in the face of reasoned defense—a defense Winsor lent his voice to in England and now again in France, lest his silence be mistaken for acquiescence.[57] He knew that gas could succeed, because it already had, and because its advocates were experts on the side of truth.

But the anti–gas lamp contingent was also quick to arrogate the mantel of enlightenment to itself. In October 1823 the Romantic authors Charles Nodier and Amédée Pichot denounced gas lamps at the behest of the oil lamp industry in their pamphlet *Essai critique sur le gaz hydrogène*.[58] They began the pamphlet with a dialogue between a medical doctor and his friend who had recently returned to Paris. The friend enumerates a list of terrible things that had happened to or near him since his return: an explosion in the Palais-Royal, a glass of tainted water at a café, a strange smell at the theater, coupled with harsh lighting that made all the women look ugly, but then extinguished in the middle of the performance, forcing the man to flee with one hand on his watch and the other on his purse. To each of his friend's complaints, the doctor replies, "*C'est le gaz hydrogène*." The doctor insists that given the dangers and inconveniences of gas lamps "those of us who live by the progress of *les lumières*, will count ourselves among their enemies." The friend accuses the doctor of using puns to make light (no pun intended!) of a very serious situation and further contends that such cavalier language takes philosophical disinterestedness into cynicism. The doctor disagrees, insisting that "the people who would accuse us of hatred for *les lumières* take this metaphor literally, and are completely convinced that the progress of human intelligence is the direct result of the intensity of the luminous source by which we are lit."[59] Is it not absurd of our opponents, Nodier and Pichot implicitly ask, to believe that brighter lamps lead to greater intellectual breakthroughs?

Nodier and Pichot were well aware that the pro-gaslight partisans would accuse them of reactionism, of "dreading the glare of the day," of "rejecting . . . all modern applications of science," of "being so far behind the times, that we can only hope for a welcome among the dead."[60] But they vehemently rejected that characterization—at one point the doctor says it would take someone "truly enlightened" (*vraiment éclairé*—lit. "truly lit") to defend the anti-gaslight position. For Nodier and Pichot, any enlightened opponent of gaslight was also a true patriot, someone who "prefers France to England, national industry to

57. Winsor, *Résumé historique et démonstratif*, 1–3.

58. This pamphlet is listed in the Oct. 11, 1823, issue of the weekly *Bibliographie de la France, ou Journal général de l'imprimerie et de la librairie*, the official list of legally published material. The oil industry's role in the genesis of the pamphlet is attested to in Benjamin, *Arcades Project*, 567.

59. Nodier and Pichot, *Essai critique sur le gaz hydrogène*, xi–xii.

60. Nodier and Pichot, *Essai critique sur le gaz hydrogène*, xiv.

foreign industry, public safety over the success of a few speculators."[61] In Nodier and Pichot's estimation, only the anti–gas lamp, and therefore pro-French, contingent could claim true enlightenment.

The idea that opposition to a new lighting technology could in itself be enlightened and patriotic may seem strange on initial examination. It is perhaps made only stranger when one remembers that Nodier and Pichot were themselves proponents of Romanticism—a new literary genre that sought to cast off the rules and strictures of French classicism—and found inspiration, at least in part, in the Romantic literature of Britain and Germany. Nodier was a member of the royalist Romantic circle to which a young Victor Hugo also belonged. Pichot wrote about and translated Walter Scott and Lord Byron. Six months after the publication of *Essai critique*, the Académie Française would publicly denounce Romanticism, in part for being un-French, but also for being culturally disruptive—for seeking to overturn well-established rules of art that had long served French culture.[62] Nodier and Pichot would have recognized the irony. At one point in their dialogue the doctor says that pro-gaslight partisans would accuse them of "not daring to do anything new, in idea or in form," to which the friend replies, "But aren't we *Romantics*?"[63]

Critics both for and against Romanticism frequently referred to it as *la nouvelle littérature*. Its novelty was one its most salient, and most debated, features. But, despite its newness, its modernity, Romanticism was also highly associated with a preoccupation with the medieval past, with gloom, with darkness, melancholy, and the gothic. The classicist writer Jean Charles Dominique de Lacretelle, expressing concern for Romanticism's foreignness as well as its melancholy, worried "that we will no longer recognize the French under the borrowed lugubrious clothing of our neighbors."[64] In 1824 the Perpetual Secretary of l'Académie Française, Louis Simon Auger, asserted that Romantics embraced the sad, the negative, the ideal, the vague, and the mysterious.[65] Many French Romantics, including Nodier and Hugo, sought to distance Romanticism from the perceived stain of association with popular gothic literature, with its fantastical elements and moody atmosphere. Nodier called this false Romanticism *l'école frénétique* and insisted those who conflated the two genres did so by assuming that all new forms of literature were Romantic. In a review of Christian Spiess's *Le petit Pierre*, a ghost story set in the thirteenth century, Nodier insisted that *l'école frénétique* was not truly novel: "They are new only once, if they even manage to

61. Nodier and Pichot, *Essai critique sur le gaz hydrogène*, xv.
62. Auger, "Discours sur le romantisme," 13–16.
63. Nodier and Pichot, *Essai critique sur le gaz hydrogène*, xiv.
64. *Annales de la littérature et des arts*, "Société royale des bonnes-lettres," 420.
65. Auger, "Discours sur le romantisme," 21.

be new once, and after that they forever walk the shrouds of their specters under the darkness of their vaults, or in the long halls of an uninhabited castle, or in the sunken chapels of a Gothic choir." He argued that these works should be judged "by the monotony of their variety, by the pale and tiresome uniformity of their colors."[66] While much of *Essai critique* focused on the same concerns over the dangers of gaslight that other anti–gas lamp partisans professed (bad smells, risk of explosion), Nodier and Pichot also made aesthetic arguments about the visual effects of gas lighting that directly parallel Nodier's criticisms of false Romanticism. In the dialogue section, when the friend attends the theater, he notices "lovely faces, lit uniformly, monochrome and flat, like cold white paper dolls . . . without color," and with no shadows to bring out "the elegant relief of their forms and the graceful suppleness of their postures."[67] Gaslight, like false Romantic writing, created a world without depth or complexity, without variation or saturation. For Nodier, advocating for Romantic literature and advocating against gas lamps constituted a unified aesthetic project.

Some scholars, like many of Nodier and Pichot's contemporaries, see anti–gas lamp sentiment as commensurate with Romanticism's more gothic tendencies. Agnès Bovet-Pavy, in *Lumières sur la ville*, argues that "opponents of gaslight, accused of obscurantism, profess a Romantic attitude," favoring shadow and the mysterious depths of night.[68] But Nodier and Pichot's aesthetic oppositions to gaslight were never about brightness per se. They even acknowledged that of all the disadvantages of gas lamps, brightness was certainly the least, and that "one could, without partiality, consider it an advantage."[69] Nodier rejected *l'école frénétique* for its darkness; why would he oppose gaslight for its brightness?

Although Romanticism has sometimes been characterized as hostile to technological or scientific development, John Tresch, in *Romantic Machine*, makes clear that an explicitly Romantic understanding of science and of machines shaped the development of French mechanical science and industry in the early nineteenth century.[70] Tiago Saraiva and Ana Cardoso de Matos similarly insist that in midnineteenth-century Lisbon technology was Romantic culture—that the gas lamps illuminating workshops "opened up the city for new nocturnal, Romantic experiences," expressed in poetry and in theater, and that Portuguese Romantics saw a continuity between the harmony of nature and the technologies of steam and

66. Nodier, *"Le petit Pierre* (deuxième article)," 182–83.
67. Nodier and Pichot, *Essai critique sur le gaz hydrogène,* ix.
68. Bovet-Pavy notes the concurrent popularity of John Field's nocturnes, Edward Young's *Night Thoughts,* and Novalis's *Hymns to the Night* with the introduction of urban gas lighting, suggesting that new forms of illumination and artistic appreciation and interpretation of the meaning of night went hand in hand (*Lumières sur la ville,* 92).
69. Nodier and Pichot, *Essai critique sur le gaz hydrogène,* 86.
70. Tresch, *Romantic Machine.*

gas.[71] In light of these insights, rather than imagining that Nodier and Pichot were rejecting gas light technology in favor of a "Romantic" vision of a pristine natural world, or a gloomy gothic one, we might instead understand their arguments against gas lighting as expressive of a Romantic theory of technology. Further drawing out the parallel between his criticisms of *l'école frénétique* and gaslight suggests that for Nodier gas lighting, like the gothic, represented a false or empty novelty—innovation for innovation's sake. In *Essai critique* Nodier and Pichot presented pro-gaslight partisans as blinded by "neomania." When the friend asks the doctor how something so popular in England found advocates in France, the doctor replies, "Like all innovations that we accept without examination . . . and abandon entirely when they lose the exciting merit of being new."[72]

Nodier and Pichot insisted that they were not proscribing this new mode of lighting without thought. Instead, they wanted to instigate an open discussion of its pros and cons to ensure that France not adopt a new and potentially dangerous technology simply because it existed and was new.[73] In their estimation, spreading "the truth" about gas lamps was what was really in the services of *les lumières*, and by rejecting the hollow novelty of illumination by gas, they opened up the possibility of embracing something robust and carefully considered—something truly modern and not just new. One review of the pamphlet argued that even if *Essai critique* could not stop the spread of gas lamps, something the reviewer deemed a foregone conclusion by 1823, it would be read with interest for its many interesting facts about the processing of coal gas and the science of combustion. "And in truth," the reviewer wrote, "there is no way to ignore those things in the century *des Lumières*."[74] We cannot ignore science and claim to be modern, he seemed to suggest. But, Nodier and Pichot believed, rejecting a new technology for sound reasons could be perfectly compatible with a vision for a modern France.

But what was modernity in Restoration France? As Marcel Roncayolo points out, defining modernity and determining when one should apply the term is difficult because "modernity" belongs more to a world of representations than to a world of facts.[75] In early nineteenth-century Paris the theater grappled directly, albeit through representation, with the battle of ancients and moderns, the conflict of revolution and counterrevolution at the heart of the Restoration. One way it did this was through a subgenre of theater built around a conflict between a postrevolutionary and an Old Regime practice, custom, or profession. These

71. Saraiva and de Matos, "Technological Nocturne," 449–53.
72. Nodier and Pichot, *Essai critique sur le gaz hydrogène*, 2, xi.
73. Nodier and Pichot, *Essai critique sur le gaz hydrogène*, 4–5.
74. V., "*Essai critique sur le gaz hydrogène*," 253.
75. Roncayolo, "La modernité?," 28.

plays, which usually feature two main characters, one modern and one traditional, frequently use marriage, or the promise thereof, as a means to resolve the
conflict at the heart of the play, a conflict that is often presented as generational.[76]
But whether or not resolution came in the form of a marriage, these plays end
either with a compromise between the old and the new or with a recognition
that it is a tension that must be lived with. In all these instances, the plays seemed
to be sublimating the anxieties at the heart of Restoration politics—how do we
build a new France when we cannot agree on whether *to restore* means a return to
the world before 1789 or not? How do we reconcile our differences and build
a new vision of monarchy fit for an increasingly modernizing world?

These plays represented Restoration conflict through a wide variety of topics, including hairstyles, the practice of law, printing, and, sure enough, gas lamps.[77]
On February 4, 1823, *Le magasin de lumière*, a play about whether new gas lamp
technology would replace oil lamps, played for the first time at the Théâtre du
Gymnase. Intended to celebrate the Gymnase's new gas lamps,[78] announcements
for the play in the theater daily the *Miroir des spectacles* were accompanied by
the declaration, "The room will be lit by gas."[79] The play focused on Legras, an
oil lamp entrepreneur; his son Isidore; Robinet, a gas lamps entrepreneur; and
Robinet's sister Estelle. Isidore and Estelle want to marry, but Legras is concerned about Isidore's ability to provide for Estelle. He does not believe that gas
lamps will take off, noting that even though business at his oil lantern shop has
declined slightly he still has plenty of clients, including many of the theaters. He
wants Robinet to prove that the gas lamp business is viable before he consents to
the marriage. Robinet admits that he has only two clients but points to its popularity in London as evidence that gaslight is the future. Legras wants more proof,
but with it he will happily consent to the union of Isidore and Estelle. Their
marriage, Legras says, would be the only way to create understanding between
"the old and the new method."[80]

Le magasin de lumière fits squarely in the genre of the theater of the old of
the new, complete with a generational conflict resolved through marriage. Robinet's sister was young enough to marry Legras's son, suggesting that Robinet was
as much as a generation younger than Legras. But while the formula of *Le magasin de lumière* would have been very familiar to French theatergoers, it should be

76. We can probably credit Eugène Scribe with this subgenre of French theater: his *L'intérieur de
l'étude, ou le procureur et l'avoué* (*Inside the Study, or the Prosecutor and the Solicitor*), cowritten with Jean-
Henri Dupin, from February 1821, appears to be one of the earliest examples.
77. See Scribe et al., *Le coiffeur et le perruquier*; Scribe and Dupin, *L'intérieur de l'étude*; and Artois
et al., *L'imprimeur sans caractère*.
78. *Almanach des spectacles pour l'an 1824*, 235.
79. See, e.g., *Miroir des spectacles*, Feb. 14, 1823.
80. Ferdinand et al., *Le magasin de lumière*, 5.

noted that in February 1823, when the play premiered, the gas lamp debate had not reached the intensity it would later that year.[81] The whole audience would have been aware of gas lamps—if only because gaslight illuminated the very theater in which they sat—but the play likely introduced some Parisians to idea of the rivalry between gas lamps and oil lanterns. Unsurprisingly, given that the theater was using the play to advertise its new gas lamps, the characters made no mention of the potential dangers of gas that would preoccupy so much of the public gas lamp debate. Instead, the play focused on the financial viability and staying power of gas lighting. Would this new technology prove to be a passing fad, or would it eventually supplant oil lanterns? What would triumph in Restoration France, the new or the old?

Much as Nodier and Pichot had hinted at the associations between lighting and literature in their pamphlet, the play raises for its audience the question of how literature and lighting would relate in a changing France. The basic plot is quite simple: Legras takes over Robinet's store for the day to see how much business he gets, and Isidore and Estelle come to the store in different disguises to convince him of the viability and respectability of gaslight. Part of that campaign involves convincing Legras that his clients are deserting oil lanterns for gas lamps, but they also go to great lengths to convince Legras that gas lamps are compatible with classicism. At the outset of the play, Legras suggests a direct correlation between oil lanterns and classicist theater. Alone in Robinet's store, Legras sings a song about how old lighting technologies were good enough for Racine, Corneille, Voltaire, Lully, and Rameau, that they lit the way for *Athalie* and *Tartuffe*, and that he hopes that gaslight would illuminate (*éclaire*) equally beautiful art. Later, Isidore, in disguise as Météore Beausoleil, tells his father he has come to discuss with Robinet his new invention: gas distilled from the finest editions of Corneille, Racine, and Molière. The product he makes is not *gaz hydrogène* but what he calls *gaz céleste*, a manifestation of the "eternal flame" that is creative genius, the "waves of light" that spread from these writers' "sublime verses."[82] Météore claims that the light he produces will be a second sun and eliminate the night. In another disguise Isidore inquires after a supply of gas for l'Institut de France, the umbrella organization that grouped the Académies, including the Académie Française. Isidore insists that anyone who supplies them with gaslight will be the savior of the institute. He draws a direct parallel between good lighting and good literature, saying that increased light leads to more reading, and better vision to more writing.[83] In the final song of the play, Legras, persuaded by the

81. Fressoz argues that the turning point came in the fall of 1823, when the French government revoked the authorization for Pauwels's gas holder ("Gas Lighting Controversy," 731).

82. Ferdinand et al., *Le magasin de lumière*, 10.

83. Ferdinand et al., *Le magasin de lumière*, 16.

subterfuge, sings that gas is preferable to "our father's torches," and Robinet sings about classicism prevailing over Romanticism, insisting that its "Germanic style" and "dark nonsense" will not "take."[84] In these scenes, the play seems to go out of its way to connect the gas lamp debate to the *bataille romantique*, to argue that there is no incompatibility between the literature of the era of Louis XIV and the vibrant new technologies of the nineteenth century, and to imply that it is Romanticism that is backward and mired in darkness. This hopeful vision of a reconciliation between the culture of the ancien régime and the technical advances of the Restoration provided one potential avenue for compromise in this new era.

But in the theater gas lighting did more than intersect with the rhetoric of modernity—it made it possible to change how plays were staged. In the eighteenth century the stage was usually lit only by footlights at the front of the stage, because lights were not strong enough to light the stage from anywhere else. This meant that only a small part of the front of the stage was well lit, and actors would jockey for space in it. It also meant that there was no way for a stage to be lit naturalistically, from above.[85] When Lavoisier wrote his 1781 reflection on how to light a theater, he remarked on the imperfection of theater lighting. He wrote that from many perspectives it was difficult to see the faces of the actors on stage and that those who were in attendance to improve their taste and literary education could not read to follow along with the text if they were seated in a part of the theater not sufficiently lit, while others had to sit in parts of the audience that were overly lit. The latter, he noted, mattered because when lights are too bright they can wash out objects, making them difficult to distinguish one from the other. Ideally, he argued, the lights on stage would be brighter than those of the audience.[86] But distinguishing between stage lighting and house lighting was not technically possible at the turn of the nineteenth century, nor was it always desired.[87] Gaslight allowed for a greater distinction between the lighting of the stage and the lighting of the audience, which made it more possible to implement the suggestions of Lavoisier and theater designers of the eighteenth century.

However, as the story from the *Diable boiteux* discussed above suggests, the implementation of this new technology had its issues. The Théâtre du Gymnase was not the first to run out of gas, and therefore light, in the middle of a performance. Both the Opéra and the Feydeau also had incidents where the

84. Ferdinand et al., *Le magasin de lumière*, 27.
85. Schivelbusch, *Disenchanted Night*, 192–94.
86. Lavoisier, "Mémoire sur la manière," 91–92.
87. Under Napoleon, for example, the whole opera house was fully lit, partly for the sake of security. Johnson, *Listening in Paris*, 167.

theater suddenly went dark. The gas lamps also went out in the middle of the August 7, 1822, performance of *Rob Roy* in Edinburgh, because the gas company had underestimated the amount needed for the evening.[88] And anxieties over the sudden loss of light in the theater played a role in anti–gas lamp rhetoric. In Nodier and Pichot's anti–gas lamp pamphlet, one of the hardships the friend endures is the sudden darkening of a theater in the middle of a performance. Not only does he not get to see the end of the play, but he also no longer feels safe and runs out of the building, escaping other patrons' cries of terror while making sure to protect his possessions so he cannot be pickpocketed.[89] In another anti–gas lamp pamphlet, the fear of sudden darkness at the theater was politicized with a glancing reference to the 1820 assassination of the duc de Berri, second in line to the throne, outside the Opéra by a man named Louvel: "Let us suppose the whole royal family at the opera, the mistake of a worker at the gasholder, and the presence of a Louvel in the crowd!!!"[90] So even as theaters took advantage of the ways gaslight was more controllable than lanterns to change the experience of theater audiences, the trade-off was they had to give up control over the supply of fuel. They had to rely on the gas companies to produce the gas in the first place and to deliver it to the gasometer both on time and in sufficient quantities.

Some critics opposed the use of gaslight in the theater on aesthetic grounds. As discussed above, one theater critic complained of the gimmicky use of gaslight to plunge the audience into total darkness, a technique he opposed not only for its inappropriateness for the scene (which took place in daylight) but also for the experience of the audience. Why, he asked, would they "keep the spectators in a darkness that tires them?"[91] Other critics worried more for the actors than the audience. One pamphlet insisted that actors at both the Feydeau and Variétés, "tired of the same brilliance that supporters of gas continue to hold good," asked for the return of Argand lamp footlights. But even after a return to the old stage lights, the pamphlet continued, actors were still forced to deal with the sore throats caused by the fumes emitted by gas lamps elsewhere in the theater.[92]

But new lighting technologies also contributed to opportunities for new forms of spectacle. And in the case of the diorama, the spectacular possibilities created by gaslight seem to have outweighed any concerns about its dangers. The diorama, invented by Louis Daguerre and Charles Bouton, was an

88. V., "*Essai critique sur le gaz hydrogène*," 249.
89. Nodier and Pichot, *Essai critique sur le gaz hydrogène*, ix, x.
90. *De l'éclairage par le gaz hydrogène*, quoted in Fressoz, "Gas Lighting Controversy," 733.
91. Le Vieil Amateur, "Spectacles," 34.
92. *De l'éclairage par le gaz hydrogène*, 6.

illusionistic theatrical experience that combined paintings on translucent canvas with lighting. An operator would change the intensity and direction of the light source to give the appearance that the paintings were changing, often from day to night but sometimes from rain to shine. Daguerre's diorama theater opened in Paris in 1822. Dioramas became so popular that in Honoré de Balzac's 1835 novel *Le Père Goriot* the artists and painters began to speak in "ramas," using words like "healthorama" and "coldorama."[93] Reviews of the diorama generally remarked on how true to life it was, how it played with light and shadow in a way that mimicked reality and transported audience members to another place. One reviewer considered the diorama so lifelike that "one believed one could hear the brook, with its murmur, tell the story of its voyage to the stones on its banks."[94] The illusion of the diorama, that time was passing or that the weather was changing, was an effective one. The "game of alternating shadow and light" that Daguerre added to his already vivid painted scenes could, in the case of his view of Canterbury Cathedral, "transport Great Britain to Paris."[95] Daguerre's diorama, a reviewer argued, because it combined not only linear perspective but also "ethereal" (*aérienne*) perspective and skillful, precise detail, produced an "image of the most exact truth" (*aspect de la plus exacte vérité*).[96] This frequent refrain that the diorama, an optical illusion made possible only by the application of an industrial lighting technology, could produce truthful representations that were faithful to nature seems somewhat odd. As Tresch points out, Daguerre's invention was in no sense natural—it was a machine. The lighting system itself moved mechanically, and the whole viewing platform rotated so viewers could experience several distinct views.[97] While there were some detractors of the diorama, who argued that artistic truth could not be found in the mechanical,[98] its incredible popularity and the overall tenor of the critical response indicate that most people disagreed.[99] This near-positive response in turn suggests that opposition to new lighting technologies could be highly contingent and that some applications of gas lighting were much more controversial than others. In theaters gaslight could be unpredictable, malodorous, and potentially dangerous, but in the diorama it opened a window to nature.

93. Balzac, *Le Père Goriot*, 45–46.
94. Brès, "Le diorama," 205.
95. *Almanach des spectacles pour l'an 1823*, 240–41.
96. *Almanach des spectacles pour l'an 1823*, 241.
97. Tresch, *Romantic Machine*, 137.
98. Tresch, *Romantic Machine*, 140.
99. One review claimed that everyone loved the diorama, for it "encouraged universal excitement, met with the approval of all artists and people of taste, and proved that in our beautiful France the love of art only needs feeding" (*Almanach des spectacles pour l'an 1823*, 240).

One thing that seems to have contributed to the level of wariness about gas lamps in any given context is the extent to which that context interacted with other concerns or controversies. As a result, the gas lamp debate illuminates a number of the political cleavages in Restoration France, including competing nationalist and cosmopolitan visions for France's future. Opponents of gaslight often emphasized its foreign origins. Nodier and Pichot remarked that it was "not very national" for foreigners to reap all the profits from the French application of a French invention, Lebon's thermolamp.[100] And as one reviewer of the Nodier and Pichot pamphlet argued, gaslight meant dependence on the British-owned Gas Light and Coke Company for the manufacture of coal gas, which "makes us tributaries of a rival, neighbor nation."[101] This concern paralleled classicist concerns about the foreignness of Romanticism. In a speech delivered to the Societé des Bonnes-Lettres in December 1823 Lacretelle insisted that embracing Romanticism would mean accepting denunciations of French classicism on the authority of people like the German playwright Friedrich Schiller, who, Lacretelle noted, denigrated Corneille, Racine, Boileau, and La Fontaine—some of France's most celebrated authors. Accepting that authority, he argued, would be tantamount to sacrificing French letters "to foreign gods." "Shall we," asked Lacretelle, lay at the Romantics' feet "the most beautiful crowns of France, and cry, like a conquered people, *Long live Germania!*?"[102] These anxieties about France's potential position as a tributary or conquered nation went beyond mere rhetorical flourish. The post-Waterloo Seventh Coalition occupation of France had ended only five years earlier, and while France's geopolitical position in Europe had strengthened by 1823—the year French forces, backed by the Quintuple Alliance, intervened in Spain to help restore King Ferdinand VII—keeping its very recent weakness at the forefront did not even require a long memory.[103] And contemporaries acknowledged this connection directly: a complaint to the Académie des Sciences compared Pauwels's gasholder to "a 'Trojan horse' left by the Englishmen after their occupation of Paris."[104]

Pro-gaslight partisans also presented their position as patriotic, although their vision of a strong France tended to emphasize prosperity and industry rather than sovereign independence. As though responding to criticisms that as an immigrant his loyalties lay elsewhere, Winsor publicly insisted on his own attachment to France and suggested that a fair assessment of his contributions to science and industry would pay no attention to nationality.[105] He also argued that gaslight

100. Nodier and Pichot, *Essai critique sur le gaz hydrogène*, 50–51.
101. V., "*Essai critique sur le gaz hydrogène*," 252.
102. *Annales de la littérature et des arts*, "Société royale des bonnes-lettres," 421.
103. See Haynes, *Our Friends the Enemies*.
104. Fressoz, "Gas Lighting Controversy," 734.
105. Winsor, *Résumé historique et démonstratif*, 3.

would be good for the French economy. Detractors who worried that a robust gas industry would depress the sale of plant oils in France,[106] he pointed out, failed to consider export markets and the many other uses for oils, as well as the jobs and profits created by the manufacture of gas, light fixtures, and other aspects of gaslight infrastructure. Moreover, he insisted, France had heretofore failed to capitalize on the value of its coal mines, something that would not be an issue with a robust coal-gas industry.[107] In a prospectus soliciting subscribers for the construction of the gasometer on Rue Faubourg-Poisonnière that helped incite the 1823 gaslight controversy, Pauwels insisted that the widespread adoption of gas lighting in England would inevitably spread to France and that the existing companies could not sufficiently meet the demand for lighting in Paris.[108] Here, England's established gaslight industry made the investment in France's own gas industry a safe choice, rather than evidence of undue foreign influence. This pro-gaslight vision of a strong France learning from its neighbors is, somewhat ironically, very close to arguments Nodier made about how French letters might benefit from familiarity with foreign works—not to slavishly imitate them but as inspiration to cast off the stifling strictures of classicist literary practices.[109]

The political meanings of gaslight were neither straightforward nor static. In the fall of 1823, when the government revoked Pauwels's license for his gasometer, the liberal press accused the royalist government of doing so because many of Pauwels's investors were liberals. The *Journal du commerce* presented this decision as government interference in commerce and accused the state of harming all those business owners who had already expended resources on retrofitting their shops and cafés with pipes and lamps that would now be useless.[110] In this moment, for these partisans, the government was the enemy of gas lighting, and government regulation was the enemy of commerce. For the anti–gas lamp contingent, the government was simply making a sound choice given the obvious dangers of gas as both an explosive and a noxious substance. Some even claimed that gas lighting was a threat to public security, because bad actors could blow up a gas holder or orchestrate a blackout to capitalize on the darkness or general confusion, potentially as part of a revolt against the government.[111] Notably, both these positions presented gas lamps as divorced from state power, and even a potential threat to that power.

In other instances, public lighting could be seen as synonymous with and symbolic of state power. Gas lamps, because they required infrastructure and

---

106. An argument made in the pamphlet *De l'éclairage par le gaz hydrogène*, 17–18.
107. Winsor, *Résumé historique et démonstratif*, 26–28.
108. *Société pour l'éclairage au gaz hydrogène inodore*.
109. Nodier, "Variétés," 2; Nodier, "*Le petit Pierre* (premier article)," 77.
110. *Journal du commerce*, Sept. 23, 1823.
111. *De l'éclairage par le gaz hydrogène*, 15, 19–21.

regulation to a greater extent than other kinds of lanterns, implied government intervention. As early as 1822 the Restoration government divided Paris into seven gas lamp districts and gave different gas companies the monopoly over the distribution of gas in each district. This was intended to avoid the issues that might arise from having more than one company lay down pipes in the same streets, and so that it would be clear which company was responsible for any explosion or leak that might occur.[112] Louis XVIII founded one of these companies, the Compagnie Royale de l'Eclairage par le Gaz, in 1818 to light the Opéra.[113] And, as noted above, in 1824 the government stepped in to create safety regulations for gas infrastructure and put the Paris police in charge of monitoring that infrastructure for compliance with the regulations. Moreover, streetlights had long been associated with increased police surveillance at night, going all the way back to Louis XIV's installation of candle lanterns in 1667.

This association of public lighting and police surveillance became particularly clear during the three days of the July Revolution. In July 1830 revolutionaries engaged in widespread lantern smashing. This was partly a tactical measure: eliminating the lights on a street made it more difficult for the government forces to root out opposition.[114] But the rhetoric of lantern smashing associated street lighting with other forms of government domination and overreach. In his description of the fighting during the July Revolution J. P. R. Cuisin wrote that "in the middle of these continually mounting horrors, night covered the capital with its shadow, and at this same moment the people began to break all the *réverbères*. . . . Along with the *réverbères*, other signs of the treacherous king were also destroyed, because we did not want to suffer the existence of a single effigy."[115] In this revolutionary moment the streetlights solidified as symbols of state power, and rhetorically even effigies of Charles X himself. Moreover, the destruction of the lights of Paris, Cuisin noted, made the city seem deserted, as if destroyed by a plague.[116] The idea that Paris could seem deserted, despite the fighting in the streets, because it was not lighted suggests that street lighting loomed large in the public's experience of Paris. By 1830, then, not only were lights, with their related infrastructure, seen as tools of the state, but they were also normalized to the extent that their destruction transformed the city, even amid the uprising that would end the Restoration.

Public lighting changed how Paris was seen and experienced at night. While gas lamps were only the newest lighting technology to light Paris's public and

112. Fressoz, "Gas Lighting Controversy," 747–48.
113. Fressoz, "Gaz, gazomètres, expertise et controverses," 68.
114. Schivelbusch, *Disenchanted Night*, 105.
115. Cuisin, *Les barricades immortelles*, 129–30.
116. Cuisin, *Les barricades immortelles*, 130.

commercial spaces, the controversy surrounding them entangled debates about lighting with debates about the future of France, its politics and its economics, and debates about new and old forms of literature. Gas lamps could be invoked as symbols of progress and modernity, or as symbols of hollow innovation for its own sake. These lights therefore took on a number of different meanings that shifted throughout the decade. Gaslight could be progressive and patriotic, it could be dangerous and foreign, it could be a tool of the state, or it could be a symbol of true knowledge illuminating the dark.

ELIZABETH DELLA ZAZZERA is assistant professor in residence of history at the University of Connecticut and associate director of communications and outreach at the UConn Humanities Institute. Her scholarship focuses on the intellectual history of material texts and urban environments in revolutionary and postrevolutionary France.

Acknowledgments

This article began as a workshop paper for the 2014–15 Penn Humanities Forum. The author would like to thank that year's fellows for their generous advice and engagement. Thanks also to Warren Breckman, Lori Daggar, Dani Holtz, and the late Jonathan Steinberg for feedback on early drafts. This article also benefited from comments by the author's colleagues in the 2020–21 University of Connecticut Humanities Institute writing group (special thanks to Helen Rozwadowski and Amanda Crawford) and those in the nineteenth-century French History Reading Group (special thanks to Hollis Clayson, Charlotte Faucher, and Charles Rearick). The author would also like to thank the anonymous reviewers and the special issue editors for their invaluable suggestions.

References

Allemagne, Henry René d'. *Histoire du luminaire depuis l'époque romaine jusqu'au XIXe siècle*. Paris, 1891.

Almanach des spectacles pour l'an 1823. Paris, 1823.

Almanach des spectacles pour l'an 1824. Paris, 1824.

Annales de la littérature et des arts. "Société royale des bonnes-lettres: Séance d'ouverture du 4 décembre." No. 166 (1823): 415–24.

Annales de la littérature et des arts. "Théâtre Royal Italien." No. 123 (1823): 189–92.

Artois, Armand d', Gabriel de Lurieu, and Francis d'Allarde. *L'imprimeur sans caractère, ou Le classique et le romantique comédie-vaudeville en un acte*. 2nd ed. Paris, 1824.

Auger, Louis Simon. "Discours sur le romantisme, prononcé dans la séance annuelle des quatre Académies du 24 avril 1824." In *Recueil factice de manifestes pro et antiromantiques*, 3–28. Slatkine Reprints, 1974.

Balzac, Honoré de. *Le Père Goriot*. Im Bertelsmann Lesering, 1962.

Benjamin, Walter. *The Arcades Project*, translated by Howard Eiland and Kevin McLaughlin. Belknap Press, 2002.

Bogard, Paul. *The End of Night: Searching for Natural Darkness in an Age of Artificial Light*. Back Bay Books, 2014.

Boutry, Philippe. "Restauration." In *Les noms d'époque: De "Restauration" à "années de plomb,"* edited by Dominique Kalifa, 27–54. Gallimard, 2020.

Bovet-Pavy, Agnès. *Lumières sur la ville: Une histoire de l'éclairage urbain.* Les Peregrines, 2018.

Brès, J. P. "Le diorama." *Annales de la littérature et des arts,* no. 98 (1822): 203–8.

Clayson, Hollis. *Illuminated Paris: Essays on Art and Lighting in the Belle Époque.* University of Chicago Press, 2019.

Cuisin, J. P. R. *Les barricades immortelles du peuple de Paris.* Paris, 1830.

Darnton, Robert. *The Great Cat Massacre and Other Episodes in French Cultural History.* Basic Books 1984.

Delattre, Simone. *Les douze heures noires: La nuit à Paris au XIXe siècle.* Albin Michel, 2000.

De l'éclairage par le gaz hydrogène. Paris, 1823.

Della Zazzera, Elizabeth. "Romanticism in Print: Periodicals and the Politics of Aesthetics in Restoration Paris." PhD diss., University of Pennsylvania, 2016.

Diable boiteux: Journal des spectacles, des moeurs et de la littérature, Oct. 2, 1823.

Diable boiteux: Journal des spectacles, des moeurs et de la littérature, Nov. 9, 1823.

Diable boiteux: Journal des spectacles, des moeurs et de la littérature. "Du gaz hydrogène et du Faubourg-Poissonnière." September 11, 1823.

Dreux du Radier, Jean-François, Jean Lebeuf, Antoine Le Camus, and François-Louis Jamet. *Essai historique, critique, philologique, politique, moral, littéraire et galant sur les lanternes, leur origine, leur forme, leur utilité, &c. &c.* Dole, 1755.

Duvergier, Jean Baptiste. *Collection complète des lois, décrets, ordonnances, réglemens et avis du Conseil d'état.* Paris, 1828.

Ferdinand, Langlé, Emmanuel Théaulon, Ramond de la Croisette, and Mathurin-Joseph Brisette. *Le magasin de lumière: Scènes à propos de l'éclairage par le gaz.* Paris, 1823.

Fressoz, Jean-Baptiste. "The Gas Lighting Controversy: Technological Risk, Expertise, and Regulation in Nineteenth-Century Paris and London." *Journal of Urban History* 33, no. 5 (2007): 729–55.

Fressoz, Jean-Baptiste. "Gaz, gazomètres, expertise et controverses: Londres, Paris, 1815–1860." In *Risques et prises de risques dans les sociétés industrielles,* edited by Denis Varaschin, 43–85. Peter Lang, 2007.

Guillerme, André. "Enclosing Nature in the City: Supplying Light and Water to Paris, 1770–1840." *Construction History* 26 (2011): 79–93.

Haynes, Christine. *Our Friends the Enemies: The Occupation of France After Napoleon.* Harvard University Press, 2018.

Hugo, Victor. *Actes et paroles: Pendant l'exil, 1852–1870.* Paris, 1875.

Hugo, Victor. *Hernani.* Paris, 1830.

Johnson, James H. *Listening in Paris: A Cultural History.* University of California Press, 1994.

Journal du commerce, September 23, 1823.

Koslofsky, Craig. *Evening's Empire: A History of the Night in Early Modern Europe.* Illustrated ed. Cambridge University Press, 2011.

Lavoisier, Antoine Laurent. "Mémoire sur la manière d'éclairer les salles de spectacles." In vol. 3 of *Oeuvres de Lavoisier: III,* 91–102. Paris, 1865.

Le Vieil Amateur. "Spectacles." *Annales de la littérature et des arts,* no. 143 (1823): 31–36.

Levitt, Theresa. *The Shadow of Enlightenment: Optical and Political Transparency, 1789–1848.* Oxford University Press, 2009.

Levitt, Theresa. *A Short Bright Flash: Augustin Fresnel and the Birth of the Modern Lighthouse.* W.W. Norton, 2013.

Loth, Arthur. *Univers,* January 29, 1876.

McMahon, Darrin M. "Illuminating the Enlightenment: Public Lighting Practices in the Siècle des Lumières." *Past and Present*, no. 240 (2018): 119–59.

Mercier, Louis-Sébastien. *Tableau de Paris*. Hamburg, 1781.

Miroir des spectacles, des moeurs, et des arts, February 14, 1823.

Nodier, Charles. "*Le petit Pierre*, traduit de l'allemand, de Speiss (premier article)." *Annales de la littérature et des arts*, no. 16 (1821): 77–83.

Nodier, Charles. "*Le petit Pierre*, traduit de l'allemand, de Speiss (deuxième article)." *Annales de la littérature et des arts*, no. 31 (1821): 175–84.

Nodier, Charles. "Variétés: *De l'Allemagne*, par Mme de Staël, cinquième edition." *Journal des débats*, Nov. 8, 1818.

Nodier, Charles, and Amédée Pichot. *Essai critique sur le gaz hydrogène et les divers modes d'éclairage artificiel*. Paris, 1823.

Pain, Joseph, and C. de Beauregard. *Nouveaux tableaux de Paris, ou Observations sur les moeurs et usages des Parisiens au commencement du XIXe siècle*. Vol. 1. Paris, 1828.

Roncayolo, Marcel. "La modernité? Approche des conceptions de la ville et de la Paris capitale . . . avant Baudelaire." In *La modernité avant Haussmann: Formes de l'espace urbain à Paris, 1801–1853*, edited by Karen Bowie, 27–37. Recherches, 2001.

Saraiva, Tiago, and Ana Cardoso de Matos. "Technological Nocturne: The Lisbon Industrial Institute and Romantic Engineering (1849–1888)." *Technology and Culture* 58, no. 2 (2017): 422–58.

Schivelbusch, Wolfgang. *Disenchanted Night: The Industrialization of Light in the Nineteenth Century*. University of California Press, 1988.

Schlör, Joachim. *Nights in the Big City: Paris, Berlin, London 1840–1930*. Reaktion Books, 2016.

Scribe, Eugène, and Jean-Henri Dupin. *L'intérieur de l'étude, ou Le procureur et l'avoué, comedie-vaudeville en un acte*. Paris, 1821.

Scribe, Eugène, Edouard Mazères, and M. Saint-Laurent. *Le coiffeur et le perruquier*. Paris, 1824.

Simond, Louis. *Voyage d'un français en Angleterre pendant les années 1810 et 1811*. Vol. 2. Paris, 1816.

Société pour l'éclairage au gaz hydrogène inodore, par le procédé de M. le chevalier Pauwels père. Imprimerie de Nouzou, 1822.

Tomory, Leslie. "Building the First Gas Network, 1812–1820." *Technology and Culture* 52, no. 1 (2011): 75–102.

Tresch, John. *The Romantic Machine: Utopian Science and Technology After Napoleon*. University of Chicago Press, 2012.

V. "*Essai critique sur le gaz hydrogène et sur les divers modes d'éclairage*, par MM Ch. Nodier et Amédé Pichot." *Annales de la littérature et des arts*, no. 162 (1823): 248–53.

Williot, J.-P. "Naissance d'un reseau gazier à Paris au XIXème siècle: Distribution gazière et éclairage." *Histoire, économie et société* 8, no. 4 (1989): 569–91.

Winsor, Frederick Albert. *Résumé historique et démonstratif sur l'éclairage par le gaz hydrogène*. Paris, 1824.

Zallen, Jeremy. *American Lucifers: The Dark History of Artificial Light, 1750–1865*. University of North Carolina Press, 2019.

Louise Michel et les savoirs de l'exil

Les traversées socio-épistémiques de la « Minerve populaire » au bagne de Nouvelle-Calédonie

VOLNY FAGES, JÉRÔME LAMY, ET FLORIAN MATHIEU

PRÉCIS   Louise Michel, condamnée au bagne après la Commune, est envoyée en Nouvelle-Calédonie de 1873 à 1880. Durant cette période, elle produit et transmet des savoirs dans trois champs d'investigation savante. D'abord, elle pratique une science naturelle mêlant connaissances académiques et formulations sensibles des savoirs. Ensuite, elle travaille à une ethnologie des populations kanaks, tiraillée entre des réflexes impérialistes et une empathie émancipatrice. Enfin, elle s'engage dans la transmission de connaissances. Dans ces trois domaines savants, Louise Michel combine une approche politique des savoirs (incluant des analogies entre le monde naturel et les sociétés humaines) et une visée académique (ne niant jamais l'importance des institutions scientifiques). Sans jamais complètement adhérer aux attendus de chacun des espaces sociaux académiques, militants et pénitentiaires, sa capacité à les traverser en conservant ses propres exigences politiques et savantes signale un positionnement original entre les savoirs de la marge et ceux des centres académiques.

MOTS CLÉS   colonies pénales, amateurs, Commune, science naturelle, ethnologie

L'histoire des sciences s'est, pour l'essentiel, bâtie en détaillant la production des savoirs académiques et légitimes. L'historiographie s'est, pendant longtemps, concentrée sur les processus de disciplinarisation[1], les méthodes d'administration de la preuve[2], les dispositifs de validation des énoncés, les efforts de démarcation des espaces socio-épistémiques[3] ou encore les tentatives de contrôle des modes de publicisation des résultats scientifiques[4]. Plus récemment, les recherches se sont ouvertes aux marges des espaces académiques. Un champ inédit d'enquête s'est formé qui pointe désormais vers l'histoire des savoirs amateurs, l'histoire des savoirs populaires et l'histoire des savoirs militants. L'amateurat est aujourd'hui étudié non plus seulement comme un ensemble de pratiques opposé

---

1. Cahan, *From Natural Philosophy to the Sciences.*
2. Licoppe, *La formation de la pratique scientifique.*
3. Gieryn, *Cultural Boundaries of Science.*
4. Baldwin, *Making "Nature."*

aux logiques professionnelles, mais dans ses dynamiques historiques propres[5] qui incluent les régimes passionnels[6], la construction des carrières[7], les formes de socialisation et d'évaluation spécifiques. L'histoire des savoirs populaires est plus ancienne[8]. Elle renvoie en fait à deux manières de considérer la production de connaissances : soit des savoirs sont élaborés pour les franges les plus populaires, soit des savoirs sont forgés par des individus ou des groupes d'individus inscrits dans les couches populaires[9]. Dans le premier cas, l'accent a notamment été mis sur la diffusion des connaissances[10] et les justifications politiques d'une édification individuelle et collective par le savoir[11]. Dans le second cas, la focale s'est resserrée sur les tentatives de conserver une autonomie dans la formation des savoirs[12], sur les logiques contraignantes qui rendent visibles ces connaissances populaires[13], ainsi que sur l'épaisse profusion de ces catégories de savoirs qui enveloppent l'ordinaire et le quotidien[14]. Ces différentes façons d'envisager les savoirs amateurs et populaires permettent d'interroger la porosité des espaces sociaux où les savoirs sont produits et circulent, et de détailler les règles des lieux de science et leurs relations avec les centralités académiques.

Dans cet article, nous nous concentrons sur le cas de Louise Michel, révolutionnaire et activiste, figure majeure de la Commune de Paris et militante politique progressivement engagée dans les combats anarchistes de la fin du XIXe siècle. D'abord institutrice, Louise Michel a participé aux soixante-douze jours de révolte du printemps 1871. Condamnée à la déportation à la fin de l'année 1871, elle est d'abord emprisonnée à Auberive avant d'être embarquée pour la Nouvelle-Calédonie en août 1873. Après sa libération, en 1880, elle multiplie les conférences, les écrits politiques et les prises de position révolutionnaires et féministes[15].

Louise Michel a toujours entretenu un rapport intense aux savoirs : en tant qu'institutrice puis comme conférencière, elle s'est donnée pour rôle de transmettre des connaissances. Rendant compte d'un de ses exposés à Paris en avril 1889, sur le suffrage universel, le journal *L'égalité* loue le « tact » et la « nature intelligente »

---

5. Guillemain et Richard, « Introduction ».
6. Charvolin, Dumain, et Roux, *Passions cognitives*.
7. Chapman, *Victorian Amateur Astronomer*.
8. Conner, *Histoire populaire des sciences*.
9. Pour cette distinction, voir Fages, *Savantes nébuleuses*, 106–12.
10. Bensaude-Vincent et Rasmussen, *La science populaire*.
11. Rancière, *La nuit des prolétaires*.
12. Secord, « Science in the Pub ».
13. Ginzburg, *Le fromage et les vers*.
14. Judde de Larivière, *L'ordinaire des savoirs*.
15. Dès les années 1860, Louise Michel s'est engagée activement pour une égalité entre les hommes et les femmes. A la suite des travaux de Sidonie Verhaeghe, il faut néanmoins souligner que la qualification de Louise Michel comme féministe est bien postérieure à cette époque (Verhaeghe, « Louise Michel, féministe »).

de celle qu'il surnomme « la " Minerve populaire " »[16]. Il est donc intéressant, dans la lignée des travaux sur l'histoire des savoirs construits depuis les marges (savantes aussi bien que politiques) de comprendre, à travers le cas Louise Michel, comment une femme révolutionnaire de la fin du XIXe siècle envisageait les pratiques savantes. Et plus globalement, quelle était sa conception du savoir ? L'envisageait-elle comme un moyen d'émancipation ? S'agissait-il d'une arme contre les dominations ?

Afin de resserrer notre objet, nous avons fait le choix de nous concentrer sur la période d'exil néo-calédonien de Louise Michel. La particularité des contraintes pénitentiaires est de mettre à nu les conditions d'existence et donc de tendre les possibilités pratiques de construction et de diffusion des connaissances. Etudier spécifiquement cette période est un moyen de faire émerger les traits les plus saillants et les plus essentiels du rapport que Louise Michel entretenait avec les savoirs.

Les lieux de privation de liberté entretiennent un rapport complexe avec la production de connaissances. Auguste Blanqui, qui a passé une grande partie de sa vie enfermé, mûrit en prison ses écrits révolutionnaires et savants[17] ; Antonio Gramsci[18] y remplit ses carnets de prison ; Germaine Tillion, pendant sa captivité à Ravensbrück, poursuit la rédaction de sa thèse d'ethnologie[19]. Louise Michel assure elle-même dans ses *Mémoires* que c'est paradoxalement en prison qu'elle a trouvé le repos nécessaire pour penser, se sentir vivre, lire, écrire, et « être un peu un être libre »[20].

Les conditions du séjour néo-calédonien de Louise Michel évoluent avec le temps. Lorsqu'elle arrive sur l'île en décembre 1873 elle est détenue, avec environ sept cents autres exilés, sur la presqu'île Ducos, à Numbo, dans une enceinte fortifiée où la circulation est possible[21]. Suite à l'évasion d'Henri Rochefort en mars 1874, les conditions de détention se durcissent—il n'est plus possible, par exemple, d'accéder à la forêt. Puis en 1875, les hommes et les femmes non mariées sont séparés. Louise Michel et cinq autres femmes sont transférées dans la baie de l'Ouest (Baie N'Gi) où elles partagent un baraquement en bois. Malgré sa mise à l'écart à l'extrémité de la presqu'île, la militante révolutionnaire peut encore y recevoir des visites et du courrier, y étudier et écrire. Mais la condition de femme de Louise Michel au bagne a très probablement constitué un stigmate entravant ses entreprises savantes—comme lorsque cet exil dans l'exil l'éloigne,

16. *L'égalité*, « Le suffrage universel », 3.
17. Blanqui, *Instructions pour une prise d'armes*.
18. Frosini, « Le travail caché du prisonnier ».
19. Lamy, « La béance ».
20. Michel, *Mémoires*, 257.
21. Salas, « Louise Michel déportée politique », 240–41.

comme on le verra plus loin, de sa serre expérimentale ou lorsqu'elle doit combattre les traits misogynes de ses compagnons d'exil masculins versés dans la science.

En reconstituant les pratiques savantes de Louise Michel en déportation, notre enquête fera émerger différents espaces sociaux que la « minerve populaire » traverse et/ou mobilise à distance : ses connaissances et relations dans l'île parmi les déportés (notamment Joannès Caton, Henri Rochefort et Henry Bauër[22] avec qui elle partage des intérêts savants), ses réseaux militants en Europe, les soutiens avec lesquels elle entretient des liens d'affection et d'admiration (comme Théodore Mauté de Fleurville, Victor Hugo, Georges Clemenceau)[23], certaines autorités savantes (comme Drouyn de Lhuys, président de la Société d'acclimatation)[24]. Louise Michel écrit beaucoup à ces multiples correspondants et, malgré la lenteur des échanges imposée par la distance à la métropole[25], parvient à les mobiliser de diverses manières (envoi de livres, échanges de spécimens, recueil d'informations, etc.) pour mener à bien un projet savant multiforme. Expériences botaniques, immersion ethnologique, déploiement pédagogique, voilà les trois principaux axes d'investissement savant de Louise Michel au bagne—ce seront aussi les trois parties de cet article.

A l'intersection d'espaces socio-épistémiques très différents, empêchée par son exil dans le Pacifique, la militante révolutionnaire met en œuvre, dans chacun de ces trois centres d'intérêt savant, une manière bien à elle de produire des connaissances. Ces conditions particulières nous permettront de saisir la façon dont Louise Michel travaille la matière savante, considère les catégories et clivages de son temps, et impose ses propres cadres d'intellection. Il s'agit donc d'interroger, dans le cas de Louise Michel, les conditions de possibilité d'un investissement savant à la fois amateur, populaire et militant qui ne rompt jamais complètement avec les ancrages académiques ni les réflexes de légitimité épistémique. Comment ce complexe de relations entre des formes si disjointes de connaissance est-il mis en tension dans l'expérience pénitentiaire de Louise Michel ?

Notre enquête s'appuie sur une recherche archivistique approfondie : la Bibliothèque nationale de France, les archives de l'Abbaye d'Auberive, les archives de la préfecture de Police de Paris, et les archives de l'Institut international d'histoire sociale d'Amsterdam (qui détient l'essentiel du fonds Louise Michel) nous ont permis de mettre au jour des archives encore inexploitées (notamment celles relatives au projet d'encyclopédie enfantine) pour comprendre ces dimensions savantes de la vie en exil de Louise Michel.

---

22. Baylac, *Louise Michel*, 129–42.
23. Narayana, « Objets trouvés ».
24. Michel, *Je vous écris de ma nuit*, lettre n°205, 202.
25. Les courriers circulent difficilement et mettent au mieux six mois pour obtenir une réponse.

Louise Michel, naturaliste de Nouvelle-Calédonie

La première voie savante explorée par Louise Michel lors de sa déportation néo-calédonienne est celle de l'étude naturaliste de son environnement immédiat. Les témoignages posthumes concernant les intérêts scientifiques de la militante révolutionnaire indiquent une forte attirance pour la botanique et plus généralement les sciences naturelles. L'éditeur de ses *Mémoires*, F. Roy, confiait qu'elle maîtrisait « à fond la botanique, l'histoire naturelle »[26]. Louise Michel tenait d'ailleurs les savoirs sur les plantes et les animaux pour essentiels dans l'éducation émancipatrice que devrait recevoir les jeunes filles : « Sous prétexte de conserver l'innocence d'une jeune fille, on la laisse rêver, dans une ignorance profonde, à des choses qui ne lui feraient nulle impression, si elles lui étaient connues par de simples questions de botaniques ou d'histoire naturelle »[27].

Lors de son exil néo-calédonien, Louise Michel fait montre d'une attention particulière aux végétaux et à la faune locale. Ses *Mémoires* conservent la trace d'une description fine de son environnement naturel. Dans un premier temps, sur la presqu'île de Ducos, à Numbo, les déportés ont la possibilité de s'enfoncer dans la forêt alentour. La militante révolutionnaire détaille la composition des lianes, couvertes de « fleurs blanches ou jaunes », que des « feuilles » de « toutes les formes possibles » ourlent « en fers de flèche comme le tarot »[28]. Elle distingue « la liane à pole d'or [qui] fleurit comme l'oranger » de « la liane fuchsia [qui] couvre les arbres environnants d'une neige de bouquets blancs pareils à des fuchsias si serrés qu'on voit à peine les feuilles » ; certaines feuilles « fragiles, transparentes » sont « couvertes d'une sorte de duvet pareil à la fleur qu'on voit sur nos prunes »[29].

La petite faune suscite également l'intérêt de Louise Michel qui s'efforce, sur le même mode contemplatif, d'en donner un détail saisissant. Ses récits oscillent entre la quête entomologique et la prose poétique. Elle confie, toujours dans ses *Mémoires*, qu'elle a « trouvé, la dernière année de [s]on séjour en Calédonie, des ricins couverts de vers, au corps nu, aux allures qui [lui] ont paru celles des bombyx » ; hésitante, elle s'interroge : « Me suis-je trompée ? Le ver à soie de ricin existe-t-il à l'état sauvage en Calédonie ? C'est ce que je vérifierai peut-être plus tard »[30]. Les ressources livresques locales manquent ; ses propres connaissances sont insuffisantes ; mais Louise Michel tente, par la menue description, de saisir au plus près les animaux qu'elle croise, de donner, par leur aspect et

---

26. Roy, « Préface de l'éditeur », vii.
27. Michel, *Mémoires*, 107–8.
28. Michel, *Mémoires*, 323.
29. Michel, *Mémoires*, 324.
30. Michel, *Mémoires*, 329.

leur forme, les indices suffisants à une confirmation ultérieure. Il arrive que le compte-rendu zoologique se mue en essai littéraire. Lorsqu'elle évoque les venues saisonnières des sauterelles de Nouvelle-Calédonie, Louise Michel se fait lyrique : « Une fois, deux fois par an quelquefois, une neige grise enveloppe la presqu'île, tourbillonnant par flocons ; on en a quelquefois plus haut que les chevilles : ce sont les sauterelles »[31] ; elle ajoute, « rien de beau comme la neige grise et tournoyante des sauterelles ; tout le ciel est pris par cette teinte uniforme ; on voit au travers le soleil tamisé par les flocons d'insectes comme à travers un crible et les flocons gris tombent, tombent toujours dans des clairs-obscurs étrangement noyés »[32]. Dans cette zoologie des formes et des sensations, le savoir naturaliste reste un horizon d'attente. Louise Michel n'entend pas céder à la pure répétition de ce qui a déjà été vu et décrit. Elle souhaite participer à la cumulativité des connaissances. Ainsi, découvrant dans les anfractuosités rocheuses « de gros vers blancs, à cornes, pareilles à celles du renne et une sorte de bourgeons noirs », elle procède à une entomologie plus pointilleuse : « J'en ai vu de tout enveloppés comme des cercueils, j'en ai vu de plus ou moins ouverts, sans surprendre si c'est la première étape de la mouche-feuille, la phyllis des naturalistes. Une seule fois j'ai vu la *mouche-fleur*, je ne crois pas qu'elle ait été encore signalée »[33].

La prose naturaliste de Louise Michel en Nouvelle-Calédonie emprunte à cette « description de la nature » dont Romain Bertrand a noté qu'elle avait constitué, avec Alexander von Humboldt et Johann Wolfgang von Goethe, « le rêve d'une "histoire naturelle" attentive à tous les êtres, sans restriction ni distinction aucune ». L'enjeu de cette représentation fine des mondes végétaux et animaux était d'exploiter les « forces combinées de la science et de la littérature pour élever la "peinture de paysage" au rang d'un savoir crucial »[34]. Cette science attentive aux surfaces du monde et à la diversité et aux identités des êtres qui les peuplent, s'est opposée, tout au long du XIXe siècle, à une vision analytique, désincarnée des êtres et des choses, dont l'anatomie, invasive et destructrice, a constitué le paradigme.

Cependant, la pratique naturaliste de Louise Michel ne se cantonne pas au strict registre de la description morphologique et sensible du monde. Elle participe d'une réflexion politique plus large dans laquelle la connaissance constitue une ressource émancipatrice centrale, et où la lutte nécessite de « s'assoiffer encore de sciences et de liberté »[35]. Observant son environnement, la militante révolutionnaire rapporte ainsi sa vision d'une « algue aux raisins violets [. . .] bien

31. Michel, *Mémoires*, 327.
32. Michel, *Mémoires*, 328.
33. Michel, *Mémoires*, 330.
34. Bertrand, *Le détail du monde*, 13.
35. Michel, *Mémoires*, 128.

vivante » qui, sur la grève néo-calédonienne, peut s'immerger dans « le flot qui reviendra » ou bien « se fai[re] terrestre, cherchant à attacher ses racines au sol »[36]. Elle note, dans une perspective évolutionniste radicale, que « c'est bien ainsi que se forment ou se développent, de la plante à l'être, des organes nouveaux suivants les milieux ». Elle ajoute, précisant la portée politique, et désabusée, de son propos : « Savons-nous nous servir de l'organe rudimentaire de la liberté, des organes rudimentaires des arts, plus ou même autant que ces fucus apprenant la vie de la terre ? Je ne le crois pas »[37]. Pour Louise Michel, l'adaptation à l'environnement est une nécessité sociale autant que politique. La militante révolutionnaire s'interroge d'ailleurs sur l'influence écologique de la colonisation européenne de l'île : « La troisième année seulement, de notre séjour à la presqu'île Ducos, nous avons vu des papillons blancs ; ces insectes sont-ils triannuels ou est-ce une nouvelle variété créée par la nouvelle nourriture apportée aux insectes par les plantes d'Europe semées à la presqu'île ? On pourra le vérifier »[38].

La pratique naturaliste de Louise Michel n'est pas uniquement descriptive— même si cette dimension occupe une grande place dans ses *Mémoires*. Elle est aussi expérimentale. Les exilés de la Commune ont le droit d'établir un jardin à proximité de leur logement. Ce maraîchage carcéral est notamment l'occasion « d'améliorer l'ordinaire »[39] des détenus. Ces petits terrains laissés aux prisonniers sont à la fois dédiés à la culture vivrière, à l'horticulture, mais aussi à l'inventaire botanique et à l'expérimentation biologique. Nous disposons de plusieurs indications sur les jardins que Louise Michel a défrichés au cours de ses déplacements en Nouvelle-Calédonie. Le premier se situait à Numbo—avant le resserrement des conditions de détention, suite à l'évasion d'Henri Rochefort en mars 1874. La militante explique qu'elle avait alors « à demi démoli » une baraque « qui était inhabitée pour en faire une serre ». L'épisode a fait grand bruit puisque ses « gardiens furent épouvantés de [s]on audace : oser toucher à un *bâtiment de l'Etat* et les déportés, eux-mêmes, [lui] trouvant pas mal *de toupet*, se demandaient ce qui [lui] en arriverait à la visite du gouverneur »[40]. Après l'évasion de Rochefort, Louise Michel et les autres femmes déportées sont exilées à la baie de l'Ouest (Baie N'Gi), à proximité de la forêt dans un baraquement en bois (fig. 1). Dans un carnet qu'elle envoie le 12 août 1875 à sa mère, ses parents, et ses « amis et amies », la militante révolutionnaire a fait un « plan de [ses] jardins

36. Michel, *Mémoires*, 326–28.
37. Michel, *Mémoires*, 327.
38. Michel, *Mémoires*, 336.
39. Clair, « " User le soleil avec la pierre ponce " », 139.
40. Michel, *Mémoires*, 303.

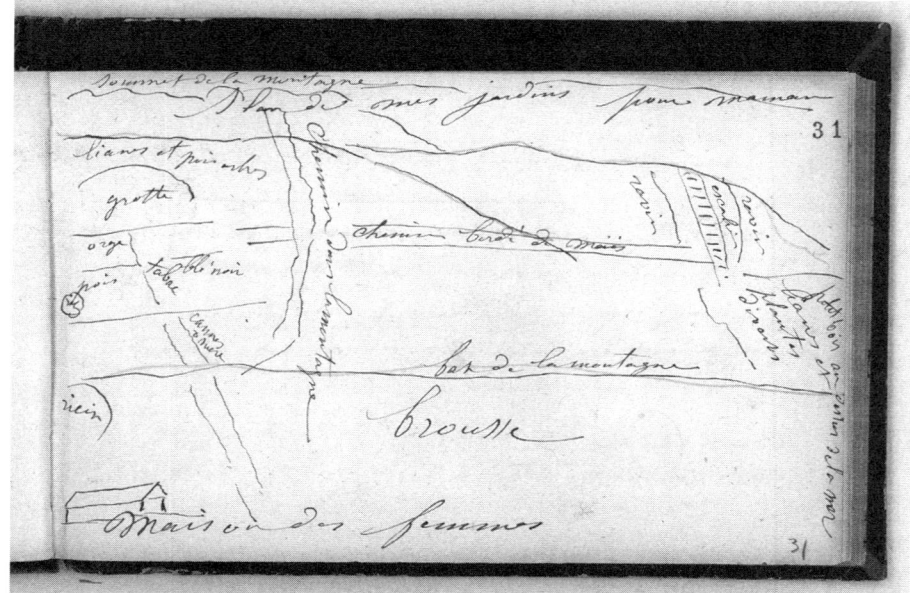

FIGURE 1 Croquis du jardin de Louise Michel, BnF, NAF 28018 (24), Carnet de Louise Michel, f°31.

pour maman » : on constate qu'elle y cultive de l'orge, des pois, du blé noir et de la canne à sucre[41].

　　Joannès Caton, un autre communard exilé, nous donne un aperçu saisissant de ce petit terrain exploité par Louise Michel. Il indique dans son journal, à la date du 9 juin 1875 : « Je l'ai trouvée à son jardin, un petit carré de moins de cent mètres carrés situé non loin de son campement, au pied de la montagne »[42]. Caton est un personnage important dans l'entourage de la militante exilée en Nouvelle-Calédonie. Il a, pendant la Commune, participé à la prise de l'hôtel de ville de Saint-Etienne. Son journal montre qu'il a un intérêt tout particulier pour la botanique. Il paraît partager cette passion avec son camarade Claude Sermet (rencontré en détention, ancien courtier en librairie, président du comité de vigilance du quatorzième arrondissement pendant la Commune de Paris). Durant les longues journées et les interminables soirées, Caton explore l'enceinte fortifiée au fil de promenades durant lesquelles il herborise. Le 1er janvier 1874, il écrit : « Je ne me lasse pas de fourrager dans les arbrisseaux et les plantes qui garnissent le talus du rivage. C'est le moment, d'ailleurs, de leur floraison et j'y fais chaque jour

41. Bibliothèque nationale de France, NAF 28018 (24), Carnet de Louise Michel, f°31.
42. Caton, *Journal d'un déporté de la Commune*, 356.

de nouvelles découvertes pour mon herbier »[43]. Il dispose, semble-t-il, de peu de moyens matériels pour mener à bien son travail de plantation[44]. Louise Michel n'a guère plus de ressources pour conduire ses expérimentations botaniques. Elle fait donc feu de tout bois pour obtenir des graines. Même si les missives peinent à circuler entre la métropole et la Nouvelle-Calédonie (la distance et la surveillance des courriers constituant de sérieux obstacles à la fluidité des échanges), il subsiste des traces archivistiques de la volonté de Louise Michel de transplanter dans sa serre océanienne des végétaux européens. Parmi celles-ci, la correspondance régulière qu'elle entretient avec Théodore Mauté de Fleurville, qui semble avoir été son référent lorsqu'elle était institutrice à Montmartre, est centrale.

Aux marges de la science académique, Mauté de Fleurville se passionne pour le magnétisme animal et a rédigé une *Physiologie élémentaire de l'agriculture*. Il est à la fois un relais des demandes de Louise Michel et une ressource pour sa documentation. Ainsi, en décembre 1878 il transmet ses écrits féministes à Clemenceau et veut lui envoyer son traité de physiologie agricole[45]. Ces échanges épistolaires savants se sont engagés avant le départ des déportés pour la Nouvelle-Calédonie. Alors que Louise Michel est encore détenue à l'Abbaye d'Auberive, Mauté de Fleurville lui confirme, le 1er août 1873, avoir remis « [sa] lettre et un exemplaire de [son] livre à Monsieur Drouhin [*sic*] de Lhuys, Président de la Société d'acclimatation qu['il] avai[t] eu l'occasion de voir autrefois ». Les deux hommes ont une « conversation toute bienveillante » à propos des futurs travaux botaniques de Louise Michel. Il la lui résume ainsi :

> 1° Il va faire part, ces jours-ci, de votre demande à la Société et vous faire envoyer à Auberive les graines que l'on pensera être utiles à la nouvelle calédonie. 2° Il vous engage à envoyer de cette île, à l'adresse ci-après, la nomenclature ou description de ce que vous aurez jugé utile à la france ou à l'algérie. Le conseil de la Société examinera et vous fera savoir ce qu'il aura décidé et il vous indiquera les moyens de lui faire parvenir vos envois[46].

En joignant Edouard Drouyn de Lhuys, Louise Michel reste en contact avec les instances académiques de la métropole. Elle a la volonté de ne pas s'éloigner démesurément des autorités épistémiques pour fournir ou obtenir des graines et envisager sa pratique botanique.

Cependant, bien d'autres voies épistolaires lui permettent de faire circuler des ressources végétales. En 1875, un rapport de police en provenance de Genève,

---

43. Caton, *Journal d'un déporté de la Commune*, 252.
44. Clair, « " User le soleil avec la pierre ponce " », 141.
45. International Institute of Social History (Amsterdam) (désormais IISHA), Louise Michel Papers, 371, Lettres de MF (Théodore Mauté de Fleurville) à Louise Michel, 16 décembre 1878.
46. IISHA, Louise Michel Papers, 370, Lettre de MF (Théodore Mauté de Fleurville) à Louise Michel, 2 août 1873.

indique que la militante révolutionnaire écrit à un correspondant genevois (« elle serait la marraine d'un de ses enfants ») pour qu'il lui envoie « de la graine »[47]. Mais la circulation botanique n'est pas à sens unique. Louise Michel fait également parvenir certains de ses végétaux vers la France. En avril 1879, elle écrit à André Léo : « Dans quelques temps je vous enverrai des graines de la grande terre »[48]. Déjà, en avril 1874, elle envoie à Mauté de Fleurville un « exemple de coton avec sa fleur » dans son courrier. Elle précise qu'il « vient à l'état sauvage à la presqu'île Ducos » et ajoute qu'« il se multiplierait autant qu'on voudrait »[49].

Les plantes ne sont pas le seul matériau scientifique que Louise Michel réclame pour ses expérimentations. Elle est également en quête de vers à soie. Dans ses *Mémoires* elle rapporte en détail ses déboires avec ses correspondants scientifiques, probablement australiens, qui ne semblent guère s'être souciés des conditions concrètes d'expédition des lépidoptères : « Pendant dix ans, j'ai demandé des œufs de ces vers ; mais (je demande aux savants qui me les ont envoyés, de raconter ceci) comme les œufs étaient dirigés sur Paris, d'où ils retournaient sur l'océan avec les lettres du courrier, ils étaient toujours éclos dans ces pérégrinations. Pourtant nous avons vu arriver des navires ayant fait relâche dans les parages d'où on m'envoyait les vers à soie »[50]. Elle a donc « bien maudit les us et les coutumes des savants qui ne font rien tout simplement »[51].

Nous savons fort peu de choses des essais botaniques de Louise Michel. L'expérience la plus connue est celle qu'elle relate dans ses *Mémoires* à propos de la vaccination des papayers. Alors qu'elle est encore à Numbo, elle s'occupe d'« arbres *en traitement* qu['elle] voulait cacher jusqu'à la complète réussite de l'essai ». Il s'agissait de « quatre papayers qu['elle] avait vaccinés au pied avec d'autres papayers malades de la jaunisse ». Elle ajoute : « Mes quatre papayers eurent la jaunisse et se rétablirent »[52]. Dans cette démarche, Louise Michel n'emprunte plus à la science des surfaces et des morphologies ; elle s'inscrit dans l'histoire des tentatives d'inoculation empirique qui ont commencé à l'époque moderne pour la variole chez les êtres humains. Le raisonnement est ici purement analogique et tend à généraliser aux plantes une hypothèse scientifique qui, au moment où elle tente ses expériences, est en train d'être validée empiriquement, pour les animaux, par Louis Pasteur. Par ailleurs, comme elle l'explique dans ses

47. Archives de la préfecture de Police de Paris, Rapport de police, Genève, 11 octobre 1875, BA 1183, dossier Louise Michel (1), document n°3662.

48. IISHA, Louise Michel Papers, 13, Lettre de Louise Michel à André Léo, 30 avril 1879.

49. Archives de l'Abbaye d'Auberive, Lettre de Louise Michel à Théodore Mauté de Fleurville, 26 avril 1874.

50. Michel, *Mémoires*, 328–29.

51. Michel, *Mémoires*, 329.

52. Michel, *Mémoires*, 303.

Mémoires, les critiques n'ont pas manqué au sein même du camp ; ce qui a renforcé la militante révolutionnaire dans sa détermination :

> J'aurais voulu réussir sur une vingtaine [d'arbres] avant d'en parler, d'autant plus que même là où tous souffraient pour la liberté, l'empire des préjugés était tel encore qu'on entendait des choses comme ceci : « S'il était vrai que la vaccine puisse s'appliquer à toutes les maladies, *la Faculté l'aurait fait ! Etes-vous docteur*, pour vous occuper de ces choses-là ? etc. » Comme si on avait à s'informer, quand une route est bonne, si c'est un âne ou un bœuf qui y est entré le premier. Jugez donc, si j'avais parlé d'étendre la vaccine aux végétaux, ce que mes *ultra universitaires* m'auraient répondu ! Il n'en est pas moins vrai qu'on essaye la vaccine de la rage, de la peste, du choléra telle que je l'avais essayée là-bas et que la sève étant du sang, on peut l'étendre jusqu'aux maladies des végétaux. En fait d'essais, si l'audace est utile, c'est surtout quand elle s'appuie sur l'analogie qui existe entre tout ce qui vit[53].

Louise Michel affronte donc, dans sa pratique botanique, ce qu'elle assimile à une forme de censure académique. Mais qui sont donc ses « ultra universitaires » ? Le journal de Joannès Caton et la correspondance de la militante permettent d'émettre une hypothèse sur la façon dont, en Nouvelle-Calédonie, les travaux expérimentaux de Louise Michel étaient reçus par ses codétenus les plus compétents dans le domaine des sciences naturelles. Précisément, alors qu'il lui rend visite seulement quatre jours après son installation à N'Gi, Joannès Caton décrit son jardin en termes peu amènes, soulignant en creux, un manque de rigueur dans les plantations et une organisation peu rationnelle de l'espace et des végétaux :

> Les sentiers y sont faits à tort et à travers et l'herbe qui les envahit ferait croire qu'ils n'existent pas si quelques plants de fraisiers plantés en bordure n'en indiquait vaguement la place. On voit des haricots plantés comme au hasard, des pois mélangés avec de la salade et divers autres légumes éparpillés en divers endroits. Partout de l'herbe, partout les plantes parasites étouffent les plantes potagères. [. . .] Elle m'y montre les rhododendrons rabougris, aux fleurs dont l'odeur est repoussante et si communs dans la montagne, la centaurée, vingt arbrisseaux sans noms parmi lesquels je reconnais seulement une *daphnée* et ce que les déportés dénomment *le faux thé*, dont toute la brousse est faite et dont plusieurs font des infusions qu'ils déclarent excellentes, mais qui ne sont qu'insipides. Louise va, vient, s'arrête, arrache ici une plante et la transplante à quelques pas, sans que l'idée d'arrangement y soit pour rien et à la façon rapide, fiévreuse dont la chose est faite je comprends pourquoi toute cette végétation paraît souffrante. Louise ne sait pas planter et elle semble croire qu'on peut impunément transplanter tout ce que l'on veut[54].

53. Michel, *Mémoires*, 304 (nous soulignons).
54. Caton, *Journal d'un déporté de la Commune*, 356–57.

Le jugement est d'autant plus sévère que si Caton est un lecteur avéré, un jardinier émérite, et un botaniste amateur chevronné[55], il n'appartient ni au monde académique ni au milieu agricole.

L'autre critique (probable) des compétences savantes de Louise Michel semble être Henri Bauër. Ce fils naturel d'Alexandre Dumas a fait des études de droit et de médecine qu'il n'a jamais terminées. Il a donc un socle de connaissances dans le domaine des sciences naturelles. Louise Michel écrit à Bauër en avril 1876 à propos d'un ouvrage que ce dernier lui a prêté. Elle reconnaît que « cette attention [l]'avait déjà étonnée », puisqu'en le « recevant », elle a « d'abord eu un instant de joie », espérant que le livre contiendrait « par les observations sur la reproduction humaine, quelque analogie avec la première production d'être vivants par les éléments en osmoses ». Mais elle se reprend, soupçonnant une forme de moquerie déguisée, une critique sous-jacente de ses méthodes : « Et puis j'ai eu souvenir de ce que disait [Gaston] Caulet [du Tayac] avec vous (ce jour où vous m'avez horripilée) par votre examen ("elle serait dans le cas de nous faire croire qu'elle lit des livres scientifiques") et j'ai cru à une plaisanterie »[56]. Indéniablement le dédain savant se double ici d'un mépris patriarcal[57]—que Louise Michel ne semble toutefois pas relever dans les échanges écrits qui nous sont parvenus, malgré sa vigilance féministe. Louise Michel pratique donc ses expériences dans un climat de moquerie et de rejet. Mais elle ne renonce pas et défend sa science des analogies.

Dans sa quête naturaliste, la militante révolutionnaire répond à tous les critères de l'amateurat tel qu'il peut être identifié à partir du XIXe siècle : ancrage local, savoirs intermédiaires (entre l'académie et les connaissances pratiques), circulations des données et des matériaux, invisibilisation du rôle des femmes[58]. Toutefois, dans ses recherches botaniques et zoologiques, Louise Michel n'est pas réductible à ces critères : elle maintient une forte tension entre ses propres pratiques et le monde académique, ne néglige pas les critiques mais conserve ses prises de positions analogistes. D'une certaine façon, elle affronte une audience multiforme depuis la Nouvelle-Calédonie, cherchant par ses expérimentations à creuser une voie socio-épistémique cohérente avec ses convictions, celle d'un rapport étroit entre la capacité biologique des êtres—tous les êtres—et leur adaptation à l'environnement.

55. Caton, *Journal d'un déporté de la Commune*, 134, 327.
56. Lettre de Louise Michel à Henri Bauër, 10 avril 1876, dans Cerf, *Le mousquetaire de la plume*, 108.
57. Toute sa vie, Louise Michel a dû faire face à de violentes attaques misogynes. Son implication politique—activité sensée être réservée aux hommes—comme sa détermination militante ont alimenté des rumeurs dégradantes à son endroit (concernant notamment sa sexualité, avec son surnom de « Vierge rouge » [Marmo Mullaney, « Sexual Politics in the Career and Legend of Louise Michel »]). Ses prétentions dans le domaine savant (quasi-exclusivement masculin encore à la fin du XIXe siècle) suscitent, même de la part de ses amis, des railleries phallocrates.
58. Guillemain et Richard, « Introduction ».

Louise Michel et les Kanaks

La seconde entreprise savante de Louise Michel durant son exil néo-calédonien est celle d'une ethnographie empathique des populations kanaks[59]. Dès sa condamnation à la déportation, la militante révolutionnaire se projette dans son futur environnement calédonien et formule le souhait d'être « utile dans ce pays nouveau »[60]. Utile à la science, on l'a vu, par l'observation botanique et zoologique, utile à l'agriculture par sa volonté de travailler à l'acclimatation d'espèces entre la France et la Nouvelle-Calédonie, mais également utile aux habitants de cette île lointaine, les Kanaks[61]. Dès son embarquement à bord de la *Virginie*, le navire qui la conduit quatre mois durant vers son exil calédonien, Louise Michel formule le double souhait de les aider à résister aux blancs et d'étudier leurs traditions et leurs modes de vie en voie de disparition. Le projet de connaissance des populations kanaks est immédiatement politique et pris dans une tension que Louise Michel ne lèvera pas : il s'inscrit dans un souci empathique porté par une volonté de participer à l'émancipation de tous les peuples, en même temps qu'il révèle une volonté de savoir ancrée dans les schèmes ethno-centrés de son temps. Dans deux lettres difficiles à dater, mais probablement écrites à bord du bateau (ou bien peu de temps après son arrivée à la presqu'île Ducos), Louise Michel explique ainsi qu'« avec les sauvages, il y aura un but doublement humanitaire d'empêcher qu'on ne les refoule par le canon en les civilisant et de faire des études historiques vraies dans les ruines »[62]. Avant même son arrivée en Nouvelle-Calédonie, Louise Michel est donc sensible au sort des Kanaks, qu'elle rapproche de son propre statut de « persécutée », et qu'elle voit comme une « race qui s'éteint »[63].

Durant toute la durée de sa déportation, elle ne se défera pas de cette vision, dominante à l'époque, plaçant les sociétés sur une échelle de développement où le degré de civilisation est fondé sur la race. Tout en développant un discours anti-impérialiste[64], Louise Michel adhère néanmoins à un modèle évolutionniste des sociétés où les races les plus avancées peuvent aider les races les plus primitives à se civiliser, reprenant ainsi des arguments classiquement utilisés pour justifier la colonisation[65]. Mais, contrairement aux autres Français vivant

---

59. Pour un aperçu général de l'histoire kanak, voir Bensa, *En pays kanak*.

60. Lettre de Mauté de Fleurville à Louise Michel, 2 août 1873 ; cf. Michel, *Je vous écris de ma nuit*, 202.

61. Nous choisissons dans ce texte d'utiliser la graphie « kanak », conformément aux volontés formulées par les peuples kanaks. La graphie « canaque » sera néanmoins utilisée dans les citations de sources primaires afin de respecter la graphie originale.

62. Lettre de Louise Michel au citoyen Paysant (et à « nos amis »), n.d. ; cf. Michel, *Je vous écris de ma nuit*, 215.

63. Lettre de Louise Michel, 11 fructidor an 81 ; cf. Michel, *Je vous écris de ma nuit*, 217.

64. Eichner, « Language of Imperialism ».

65. Bullard, *Exile to Paradise*.

en Nouvelle-Calédonie, y compris ses camarades déportés, Louise Michel mani-feste un intérêt profond pour les Kanaks, elle souhaite les comprendre et agir à leurs côtés[66].

Les interactions que Louise Michel a pu avoir avec les Kanaks sont diffi-ciles à reconstituer car les archives manquent. Autour de 1873, les contacts entre les déportés à la presqu'île Ducos et les Kanaks sont rares. Aucune tribu ne sem-ble s'être établie jusqu'alors de façon permanente sur cette presqu'île au climat aride et ne possédant que peu de cocotiers. Et, à partir de la mise en place de l'enceinte fortifiée, les autorités françaises en interdisent l'accès aux Kanaks[67]. Durant ses premiers mois de détention, Louise Michel a donc peu l'occasion de rencontrer ces autochtones qu'elle désire tant découvrir, étudier et éduquer. Peu après son arrivée, elle rencontre néanmoins un Kanak occidentalisé, parlant et lisant le français, travaillant au service d'un cantinier vendant des provisions aux déportés : Daoumi. Originaire de l'île de Lifou, fils de chef, il sera le principal infor-mateur de Louise Michel, qui durant huit mois discute régulièrement avec lui, jus-qu'en août 1874[68]. Daoumi lui chante des chansons, lui raconte des mythes, lui apprend quelques rudiments de vocabulaire des différentes tribus. Convaincue de la nécessité de sauver ce qui peut encore l'être de cette culture menacée d'extinc-tion, Louise Michel recueille cette parole et conçoit le projet de la transcrire et de la publier. « Ne pourrait-on saisir ces dialectes, étudier cette race, avant que l'ombre recouvre des choses historiquement curieuses », écrit-elle[69].

Après le 5 juin 1875, cantonnée dans la baie N'Gi, Louise Michel se concen-tre sur la tenue de son jardin et l'écriture. Elle en profite pour rédiger ses *Légendes canaques*. Ce texte rassemble quatorze « légendes » écrites dans un style s'effor-çant de reproduire leur oralité[70]. Par des phrases courtes et incisives, de fré-quentes répétitions, des ellipses temporelles, Louise Michel s'applique à restituer la voix des conteurs kanaks en mettant en avant la dimension poétique de cette oralité[71]. Comme l'explique Carolyn J. Eichner, en publiant ces légendes kanaks Louise Michel s'adresse en premier lieu à un lectorat européen, et plus spécifiquement français[72]. L'objectif de ce texte est triple. Par cette publication, il s'agit d'abord de valoriser et de faire connaître cette culture lointaine. Ensuite, l'enjeu est de garder la trace d'une tradition purement orale risquant de dispa-raître. Les travaux de Kathleen Hart ont bien mis en évidence l'aisance orale de

66. Eichner, *Feminism's Empire*, 117.

67. Dauphiné, *La déportation de Louise Michel*, 55.

68. Le 19 août 1874, après l'évasion de Rochefort, les cantiniers ne sont plus admis sur la presqu'île et Louise Michel ne reverra plus Daoumi.

69. Michel, *Légendes*, 57.

70. On consultera avec profit l'édition des *Légendes et chansons de gestes canaques* établie par Fran-çois Bogliolo (Michel, *Légendes*).

71. Soula, *Histoire littéraire de la Nouvelle-Calédonie*, 38.

72. Eichner, « Language of Imperialism », 378.

Louise Michel, qui est une grande oratrice, et l'importance de l'oralité dans les cultures populaires, notamment pour les populations les plus dominées[73]. Mais une troisième dimension la rattache à son propre agenda politique révolutionnaire. Par ce texte, Louise Michel entend en effet saper, par la valorisation des Kanaks, l'hégémonie et l'impérialisme européens afin de promouvoir une humanité unifiée et égalitaire[74]. Elle s'efforce ainsi de changer l'image que les Européens se font des Kanaks, soulignant que « la race canaque est meilleure qu'on ne le croit »[75], atténuant les écarts culturels qui auraient contribué à construire une altérité fondamentale avec les « sauvages » calédoniens. Le cas de l'anthropophagie est, sur ce point, révélateur. Les pratiques anthropophages sont en effet fréquemment mises en avant par les Français pour étayer la construction d'une image des Kanaks comme un ensemble de tribus appartenant encore à « l'âge de pierre » et se situant au plus bas de l'échelle de l'évolution des races humaines[76]. Louise Michel ne nie pas ces pratiques, mais, par le truchement des mythes qu'elle transcrit, elle inscrit le cannibalisme dans un temps ancien, et réduit cette pratique à des cas exceptionnels de vengeance ou de famine. « Il y eut depuis bien des famines et bien des guerres. Ne croyez pas que les tayos sans la faim ou sans la colère aient jamais mordu la chair humaine »[77].

La publication de ces « légendes » en 1875 ne marque pas la fin de l'intérêt que Louise Michel porte aux Kanaks et à leur culture, mais plutôt l'amorce d'un désir d'approfondissement de l'étude de ces peuples. Consciente que sa connaissance des habitants de l'archipel est encore superficielle, elle écrit en 1876 deux lettres à Victor Hugo dans lesquelles elle lui fait part de son souhait « d'aller étudier de près une tribu canaque »[78]. Elle devra attendre encore cinq ans avant de pouvoir circuler sur la Grande Terre, mais elle formule déjà le « rêve » de « passer quelques temps à Bourail » et d'« aller également (pendant assez longtemps pour [se] rendre compte de la langue et des usages) dans une tribu canaque où notre influence ne se serait pas encore fait sentir »[79]. Il s'agirait de « résider dans une tribu [...] au moins pendant un an afin de voir les coutumes et cérémonies, etc. complètement »[80].

Louise Michel ne s'établira finalement pas dans une tribu comme elle l'espérait. Durant la dernière période de sa déportation, à partir de 1879 elle est autorisée à vivre et enseigner à Nouméa. Elle y rencontre de nouveaux informateurs, originaires

73. Hart, « Oral Culture and Anticolonialism ».
74. Eichner, « Language of Imperialism », 389–90.
75. Michel, *Légendes*, 57.
76. Bullard, *Exile to Paradise*, 30–38.
77. Michel, *Légendes*, 125. « Tayo » signifie « ami » dans le lexique kanak de Louise Michel. Elle désigne par ce terme générique tous les habitants autochtones de Nouvelle-Calédonie.
78. Lettre de Louise Michel à Victor Hugo, 3 août 1876 ; cf. Michel, *Je vous écris de ma nuit*, 240.
79. Lettre de Louise Michel à Victor Hugo, 18 juin 1876 ; cf. Michel, *Je vous écris de ma nuit*, 238.
80. Lettre de Louise Michel à Victor Hugo, 3 août 1876 ; cf. Michel, *Je vous écris de ma nuit*, 240.

de la Grande Terre mais également des îles alentour (Lifou, Maré, Ouvéa, île des Pins), recueille de nouvelles légendes, et approfondit son étude des langues kanaks. Dans une lettre adressée à Mauté de Fleurville et conservée aux archives de l'Abbaye d'Auberive, elle copie par exemple un « Vocabulaire de l'île des pins d'après le déporté Bouquemont »[81] comprenant plus de 450 mots, dans l'ordre alphabétique. A Nouméa, Louise Michel rencontre également Charles Malato, un jeune homme déporté à l'âge de treize ans avec ses parents communards. Malato vient alors de rentrer de deux années passées en brousse en charge de la gestion du télégraphe à Oubatche, dans le nord-est de la Grande Terre, où il a également recueilli des récits auprès d'autres tribus kanaks[82].

Tous ces éléments constitueront le substrat d'une seconde édition des *Légendes*, en 1885, augmentée et très largement remaniée par Louise Michel. En effet, après son retour en Europe en 1880, la volonté de la militante révolution-naire d'œuvrer en faveur des Kanaks ne faiblit pas. Alors qu'elle est, une fois encore, mise en prison, cette fois-ci à la centrale de Clermont, dans l'Oise, elle établit cette nouvelle édition[83]. Celle-ci comprend maintenant vingt-quatre chapitres, dont un « conte du rat et du poulpe », « remarquable parmi ceux que Charles Malato recueillit sur la grande terre »[84]. Cette édition de 1885 se termine également par des « Fragments de vocabulaire. Mots répandus dans les tribus », et des « Numérations diverses ».

L'étude des langues kanaks constitue, pour Louise Michel, dès son arrivée, un passage obligé pour la compréhension des « sauvages » avec qui elle veut entrer en contact et saisir les coutumes. Dans l'édition de 1875 des *Légendes*, elle s'inter-roge rhétoriquement : « Le vocabulaire d'une peuplade n'est-ce pas ses mœurs, son histoire, sa physionomie ? »[85]. Et Mauté de Fleurville l'encourage en ce sens. En 1878, par exemple, il lui demande des « nouvelles de [son] dictionnaire canaque qui [lui] facilitera la grammaire à faire ensuite »[86]. Louise Michel ne publiera ni dictionnaire ni grammaire, mais uniquement ces « fragments de vocabulaire », rassemblant des mots utilisés par plusieurs tribus provenant de divers lieux de l'archipel.

On retrouve dans la manière qu'a Louise Michel d'étudier les langues kanaks le rapport sensible à son environnement qui caractérise également, on l'a

81. Archives de l'Abbaye d'Auberive, lettre de Louise Michel à Mauté de Fleurville, n.d., p. 1. Charles Bou-quemont, né en 1834, passementier, communard, est condamné à la déportation en enceinte fortifiée en Nouvelle-Calédonie. Mais sa peine est commuée le 1er août 1872 en déportation simple et il arrive à Nouméa le 9 février 1873. Il a sans doute rejoint l'île des Pins, comme beaucoup de communards condamnés à la déportation simple.

82. Dauphiné, *La déportation de Louise Michel*, 88.

83. Dauphiné, « Présentation », dans Michel, *Légendes*, 86–87.

84. Michel, *Légendes*, 149.

85. Michel, *Légendes*, 57.

86. Lettre de Mauté de Fleurville à Louise Michel, 16 décembre 1878 ; cf. Michel, *Je vous écris de ma nuit*, 245–46.

vu, son étude de la nature calédonienne. Pour le vocabulaire de l'île des Pins, sa sélection de mots est directement liée à l'observation de la nature contenant « la liste presque complète des plantes et des animaux calédoniens », dont elle précise immédiatement, comme pour en souligner le caractère amical, qu'ils sont « tous sans venin »[87].

Partant de discussions avec les quelques Kanaks avec qui elle a pu échanger, Louise Michel est également confrontée à la grande richesse et à la complexité linguistiques de l'archipel, où plusieurs dizaines de langues vernaculaires se côtoient[88]. Les « vocabulaires » de 1885 identifient des mots propres aux langues de lieux spécifiques. Mais Louise Michel s'intéresse également au pidgin véhiculaire permettant aux Kanaks de différentes tribus de communiquer entre eux, le « bichelamar »[89]. Mêlant des mots de langues kanaks, d'anglais, de français, de dialectes polynésiens et d'Asie du sud-est, le « bichelamar » est, pour la militante révolutionnaire, une « langue universelle » au même titre que les langues anciennes ou, même si elle ne le mentionne pas, au volapük créé en 1879 et à l'espéranto qui émergera peu de temps après en 1887 : « Vos philosophes discutent la possibilité d'une langue universelle choisie parmi les langues mortes, nos peuplades de l'âge de pierre font et *vivent* cette langue, en prenant chez les Anglais, les Français, les Espagnols, les Chinois, pêcheurs de Trépang, leurs mots d'usage, et en leur donnant des leurs »[90].

Louise Michel se risque ensuite à quelques analyses du « bichelamar », en s'appuyant ici encore davantage sur l'identification d'analogies—de sonorités—que sur des travaux académiques de linguistique. Elle suggère par exemple que les mots « piquinini » (enfant), « nemo » (femme) ou « popinée » (femme ou objet d'utilité) auraient une origine italienne. Ou s'étonne de retrouver la même racine chez « le *Thoth* égyptien, le *Teutatès* gaulois, le *Théos* grec, le *Tabbé* (magicien) samoyède, le *Takata*, médecin sorcier canaque, *Théo*, le tonnerre canaque, *Théama*, chef suprême des tribus »[91], dans une démarche ici encore plus morphologique qu'analytique. Etudiant les langues en autodidacte, la militante révolutionnaire adopte une vision de leur origine conforme à son positionnement politique anarchiste internationaliste. Par la recherche d'analogies entre les langues, elle cherche à renforcer la thèse d'une origine commune des langues et des races. Soulignant les ponts entre les différents peuples de l'humanité, elle se rattache à la vision évolutionniste défendue par Charles Darwin dans *The*

87. Michel, *Légendes*, 186.
88. Maurice Leenhardt dénombrait trente-sept langues et dialectes parlés en Nouvelle-Calédonie dans son ouvrage classique de 1946 (Leenhardt, *Langues et dialectes*).
89. Eichner, *Feminism's Empire*, 145–54.
90. Michel, *Légendes*, 95.
91. Michel, *Légendes*, 98.

Descent of Man en 1871 et contre les philologies dominantes de l'époque qui tendent alors à associer races et langues pour les naturaliser et les hiérarchiser[92].

A la toute fin de l'édition de 1885 des légendes, Louise Michel rassemble onze « numérations » utilisées dans l'archipel, à savoir dix façons de compter jusqu'à dix et une numération allant jusqu'à quatre-vingt-dix. Le recueil de ces numérations semble plus délicat que celui des vocabulaires. Dans une lettre à Mauté de Fleurville, celle où elle reproduit le vocabulaire de Charles Bouquemont, elle indique deux « numérotations », soulignant que la seconde est « incomplète par la difficulté de rencontrer un kanak savant »[93]. Louise Michel ne semble pas avoir réussi à compléter cette seconde numérotation, « d'après un naturel de l'île des pins », car elle ne sera pas publiée en 1885. Mais ce passage de la lettre la montre en pleine enquête ethnomathématique. Après avoir énuméré les premiers nombres :

1. Ta
2. Bat
3. Badcygne
4. Benknet
5. Bruwtonet
6. nota
7. donada
———
12. seydo

Elle ajoute ensuite : « de 7 à 12 les nombres me manquent mais d'après la relation de 7 donada à 12 seydo j'espère les reconstruire quand j'en pourrai encore avoir deux ou trois »[94].

Cette attention fine à l'identification et à la compréhension des systèmes de numération, que l'on peut probablement rapprocher de son activité d'institutrice, se retrouve également dans son texte de 1885 : « Nous avons toujours remarqué que les Canaques en nombrant avancent d'abord une main puis l'autre, puis un pied et ensuite l'autre ; leurs numérations ont presque toutes un temps d'arrêt, et quelques-unes un changement entre ces quatre séries »[95]. On voit, ici encore, l'importance de l'observation dans les pratiques savantes de Louise Michel, une observation s'appuyant sur une sincère attention à l'autre.

92. Eichner, *Feminism's Empire*, 147.
93. Archives de l'Abbaye d'Auberive, lettre de Louise Michel à Mauté de Fleurville, Vocabulaire de l'île des Pins, ca. 1875–80, 4. « Savant » signifie ici pour Louise Michel sachant compter ou dénombrer, dans une acception minimaliste, conformément au rapport d'infantilisation qu'elle entretient avec les Kanaks.
94. Archives de l'Abbaye d'Auberive, lettre de Louise Michel à Mauté de Fleurville, Vocabulaire de l'île des Pins, ca. 1875–80, 4.
95. Michel, *Légendes*, 184.

Dans le cas de ces travaux linguistiques sur les Kanaks, Louise Michel ne s'attribue pas pleinement un rôle de scientifique, qui chercherait par exemple à construire des classifications systématiques, mais plutôt celui consistant à collecter un corpus d'informations sur une culture en péril. Charge ensuite à la science académique de « s'emparer des vocabulaires, des numérations, [de] saisir sur le vif les mœurs de l'âge de pierre » pour trouver « au fond quelque chose du passé »[96]. S'appuyant ici encore sur sa vision évolutionniste des races, l'éthnomathématique de Louise Michel se change en paléomathématique, où l'étude des Kanaks nous éclairerait sur notre passé lointain.

Plongée dans un environnement dont elle ignore tout à son arrivée, obligée de se plier aux contraintes de l'enceinte fortifiée qui limitent ses mouvements, ses lectures, et la possibilité de rencontre avec les Kanaks, Louise Michel braconne sur les territoires de l'ethnographie. Malgré le souhait formulé à bord de la *Virginie*, elle ne semble pas avoir entretenu de correspondance avec la Société de géographie[97], qui aurait pu lui donner des indications méthodologiques pour consigner les traditions, usages et langues kanaks. Elle se construit donc sa propre manière de produire des connaissances sur les habitants de l'archipel, mêlant empathie, implication affective et engagement politique à un désir d'authenticité et de restitution du réel. Epistémologiquement, Louise Michel est ici prise dans une tension qu'elle ne résout pas. D'une part, conformément à son engagement permanent dans l'action (politique, sociale, amicale . . .) et à sa valorisation de l'observation sensible, la militante révolutionnaire privilégie le terrain aux spéculations. Dans l'édition de 1875 des *Légendes*, elle dénonce les « quelques voyageurs » qui ont écrit sur la Nouvelle-Calédonie « des romans auxquels on a cru tant qu'on n'y est pas venu voir »[98]. Et elle formule à plusieurs reprises, on l'a vu, le souhait d'aller vivre une année entière auprès d'une tribu kanak, valorisant ainsi les informations recueillies au plus près des autochtones, sans intermédiaire. Mais, d'autre part, son implication affective et politique la conduit à mêler le discours de ses informateurs au sien. Jouant elle-même un rôle de médiation avec les Européens, Louise Michel interprète, commente et complète les récits qu'elle transcrit : « Ces récits et chants sont ceux qui bercent toute l'humanité à son premier âge ; c'est pourquoi il est souvent facile de saisir la pensée du Canaque et de compléter la phrase »[99]. On voit poindre dans cette citation une seconde tension dans le positionnement de Louise Michel à l'égard des Kanaks, tension qui se retrouve dans ses textes à vocation ethnographique.

96. Michel, *Légendes*, 185.
97. Aucune trace de correspondance de Louise Michel avec la Société de géographie n'a été trouvée. Nous avons consulté Fierro, *Inventaire des manuscrits* ; Méhaud, *Manuscrit de la Société de géographie* ; Lemosof, *Table des matières* ; et De Martonne, *Tables générales de « La Géographie ».*
98. Michel, *Légendes*, 55.
99. Michel, *Légendes*, 56.

Les *Légendes*, dans leurs deux éditions, montrent, on l'a vu, la prégnance chez leur autrice d'une vision raciste évolutionniste et impérialiste majoritaire en France à l'époque, en particulier dans l'infantilisation constante des peuples kanaks. Mais, simultanément, Louise Michel reste fidèle à son engagement de militante anarchiste et défend une vision unifiée de l'humanité. Il est ainsi naturel pour elle de lutter pour l'émancipation des Kanaks au même titre que pour celle des peuples d'Europe[100].

En 1878, la grande révolte kanak éclate. Tout en condamnant les assassinats de blancs en brousse, dont la presse se fait très largement l'écho à Nouméa, Louise Michel semble avoir été l'une des rares déportés à avoir soutenu la révolte. En 1886, dans ses *Mémoires*, se souvenant de cette guerre, elle s'insurge contre « la supériorité qui ne se manifeste que par la destruction ! »[101].

Menant de front une ethnologie curieuse, nimbée d'empathie, et un objectif politique d'émancipation à l'endroit des Kanaks, Louise Michel n'en reste pas moins travaillée par les grands mouvements socio-épistémiques de son temps. Elle n'échappe ni à l'européanocentrisme, ni à la visée raciste et impérialiste de la mise au jour de traditions autochtones immédiatement inscrites dans une grille de lecture évolutionniste. Cette tension irrésolue entre désir émancipateur et normativités culturelles fait de Louise Michel une ethnologue engagée et singulière des rites et des coutumes kanaks. Singulière dans son positionnement socio-épistémique, Louise Michel l'est aussi dans son rapport aux institutions savantes et dans sa méthode. La militante révolutionnaire se situe à la fois à distance et en interaction avec les autorités épistémiques de la discipline naissante, et continue d'emprunter à un registre analogique ouvertement assumé (comme dans le cas de l'étude du « bichelamar »).

Louise Michel pédagogue

La troisième voie savante suivie par Louise Michel en exil est celle de la transmission *lato sensu* des connaissances. Bien que célébrée dans la mémoire militante comme éducatrice des enfants du peuple en raison de son métier d'institutrice, Louise Michel n'a en réalité laissé que peu de traces matérielles de ses activités d'enseignement[102]. Comme le montre Carolyn Eichner, sa déportation au bagne constitue une période charnière dans son parcours entre ses pratiques pédagogiques républicaines laïques d'avant la Commune, et son investissement en 1890 dans l'International School à Londres, d'inspiration profondément anarchiste[103].

100. Michel, *Légendes*, 64.
101. Michel, *Mémoires*, 359.
102. Michel, *Mémoires*, 30, 51, 150–51.
103. Eichner, *Feminism's Empire*, 154–63.

En Nouvelle-Calédonie, la militante révolutionnaire exprime dès 1874 son envie d'instruire[104], mais c'est essentiellement dans les dernières années de son séjour qu'on trouve mention de la reprise de son métier. Elle évoque d'abord dans certaines lettres à partir de 1878 sa volonté de fonder une école dans une tribu kanak[105], puis est finalement autorisée par l'administration pénitentiaire à enseigner à partir de 1879 mais à un groupe d'enfants de bagnards[106]. Elle est enfin officiellement chargée des cours de musique et de dessin dans une école élémentaire de jeunes filles à Nouméa à partir de 1880[107]. La militante révolutionnaire mentionne aussi dans ses mémoires la mise en place d'un « cours canaque » du dimanche pour adultes[108]. Si plusieurs documents attestent de son exercice d'enseignement auprès d'enfants, les traces de ce cours à destination des Kanaks sont bien plus ténues. Il n'empêche que Louise Michel a au moins envisagé d'enseigner aux Kanaks, sans qu'elle n'ait cherché, semble-t-il, à distinguer ce public de celui de ses ordinaires classes élémentaires.

En dehors de ces quelques traces déjà bien identifiées par l'historiographie, d'autres documents attestent de l'élaboration d'un projet pédagogique bien plus vaste, envisagé comme une pratique d'émancipation, qui peut être mis en relation avec ses pratiques savantes dans les domaines des sciences naturelles et de l'anthropologie précédemment présentées.

Lorsque Louise Michel commence à envisager la possibilité de fonder une école, les livres se situent au cœur de son projet[109]. Elle se retrouve dès lors confrontée aux mêmes difficultés matérielles que lors de ses demandes les années précédentes quand elle cherchait à obtenir des ouvrages de botanique ou de sciences naturelles. De la même façon, elle mobilise ainsi très probablement son réseau en métropole, dans le but de constituer une bibliothèque généraliste. Mais elle ne se contente cependant pas de faire appel à des intermédiaires « savants » (tels que Mauté de Fleurville ou Clemenceau), et s'appuie également sur sa sociabilité militante. Elle peut notamment compter sur la presse républicaine et socialiste, à l'instar du journal *Le prolétaire*[110] qui relaie ses demandes à partir de 1879, lorsqu'elle commence effectivement à enseigner : « Nous faisons un appel très pressant aux amis qui seraient heureux de faire parvenir des livres et de la musique à la vaillante citoyenne Louise Michel, pour l'aider dans la tâche ardue qu'elle a entreprise : fonder une école à la Nouvelle-Calédonie. [. . .] Nous

104. Archives de la préfecture de Police de Paris, Rapport de police, Genève, 11 octobre 1875, BA 1183, dossier Louise Michel (1), document n°3662.

105. Louise Michel, *Lettre datée du 22 octobre 1878*, dossier « lettres inédites » reproduites par Claude Rétat dans l'édition de 2021 des *Mémoires* de Louise Michel.

106. Dauphiné, *La déportation de Louise Michel*, 85–87.

107. Dauphiné, *La déportation de Louise Michel*, 93–94.

108. Michel, *Mémoires*, 357.

109. Louise Michel, *Lettre datée du 24 octobre 1878*, dossier « lettres inédites ».

110. *Le prolétaire, journal républicain des ouvriers démocrates socialistes*, Paris, 1878–84.

invitons les citoyens qui voudraient faire don de quelques livres d'instruction morale et scientifique, à le faire sans retard »[111].

On constate ainsi que le type d'ouvrages demandés dépasse le cadre des seuls enseignements de musique et de dessin dont elle était officiellement chargée (bien que ces derniers y soient inclus). La consultation du *Prolétaire* permet également de retrouver la trace d'un des donateurs ayant répondu à cet appel : un certain Jean-Baptiste Davagnier[112].

Ouvrier cambreur et secrétaire de la chambre syndicale des cuirs et peaux[113], Davagnier présente un profil assez classique de l'ouvrier militant parisien[114]. Nous ne connaissons pas la nature exacte des ouvrages envoyés par Davagnier à Louise Michel, mais nous savons que dans les semaines qui suivent son envoi de livres, Davagnier s'engage également dans le soutien matériel aux premiers amnistiés de retour en métropole[115]. Outre cette solidarité concrète avec les communard·e·s condamné·e·s, il participe au développement du mouvement socialiste en souscrivant régulièrement à un fond « Pour le développement du *Prolétaire* et la création d'autres organes ouvriers »[116]. Parallèlement à cet engagement militant, Davagnier s'illustre également dans le domaine littéraire. Déjà auteur en 1848 d'un *Réveil du peuple*[117] exaltant la veille politique révolutionnaire[118] puis de diverses chansons républicaines[119], Jean-Baptiste Davagnier publie en 1879 un *Carnaval de nos jours*[120], pièce satirique en vers dans laquelle transparaissent également ses idées politiques et sociales. Cette œuvre littéraire nous permet ainsi de rapprocher Davagnier des poètes ouvriers si bien étudiés par Jacques Rancière dans sa *Nuit des prolétaires*[121]. Bien que Davagnier soit le seul donateur ouvrier dont les archives nous ont permis de retracer le parcours, les informations rassemblées autour de son cas interrogent sur l'éventuelle existence d'un réseau de prolétaires intellectuels partageant les mêmes idéaux que Louise Michel d'émancipation par le savoir, et ayant ainsi contribué à l'envoi d'ouvrages à cette dernière au bagne, dans le cadre de la reprise de son métier d'institutrice à Nouméa à partir de 1879.

Conservé à l'International Institute of Social History (IISHA), un manuscrit de Louise Michel rédigé à Nouméa en 1879–80 révèle que ses demandes

111. *Le prolétaire*, 3 mai 1879, 8.
112. *Le prolétaire*, 29 mars 1879, 7.
113. *Le prolétaire*, 16 octobre 1880, 8.
114. Rustenholz, *Paris ouvrier*.
115. *Le prolétaire*, 10 mai 1879, 4.
116. *Le prolétaire*, 9 août 1879, 2.
117. Davagnier, *Le réveil du peuple*.
118. Pour une étude détaillée de ce motif littéraire, voir notamment Panziera, « L'exaltation de la veille politique ».
119. Davagnier, *Le bon vieux temps*.
120. Davagnier, *Le carnaval de nos jours*.
121. Rancière, *La nuit des prolétaires*.

d'ouvrages auprès de ses soutiens en métropole n'étaient pas uniquement desti- nées à alimenter une bibliothèque scolaire, mais s'inscrivaient également dans un projet d'écriture d'une encyclopédie à destination des enfants[122]. Louise Michel évoque ce projet à plusieurs reprises dans ses *Mémoires*, et finira par publier bien après son retour deux courtes brochures dans lesquelles elle ne fait qu'esquisser ses plans[123].

Le manuscrit de Nouméa, bien qu'incomplet, se compose de quatre-vingts pages de texte accompagnées de quelques illustrations. Intitulé « Encyclopédie enfantine », il permet d'appréhender les ambitions à la fois pédagogiques et d'émancipation par le savoir de Louise Michel, dans les conditions particu- lières de la fin de son séjour en Nouvelle-Calédonie. Avant d'exposer au long des pages avec plus ou moins de détails des connaissances sur diverses disciplines, l'encyclopédie s'ouvre par un premier chapitre intitulé « Géologie enfantine, vue générale ». Loin de se limiter strictement à la géologie, ce chapitre fait en réalité office d'introduction dans laquelle Louise Michel explicite son rapport au savoir : « Oui il faut que l'homme se développe il faut qu'il sache. Qu'il épelle les sciences dont nous connaissons à peine l'alphabet—une autre race lira à livre ouvert, et pourquoi pas la nôtre ? [. . .] Quand les rivages changeront, faites enfants que l'ont ait été la race humaine, après que nos aïeux aient été l'animal humain, et que nous soyons, nous, le troupeau humain »[124].

Par le biais d'une analogie animalière, la « race » selon Louise Michel est donc le stade suprême de l'évolution de l'humanité après l'« animal » puis le « troupeau »—correspondant au stade actuel de la société[125]—et les sciences se trouvent ainsi placées au cœur du projet d'émancipation.

Dans les pages qui suivent, c'est précisément la relation analogique au monde qui semble caractériser l'encyclopédisme défendu par Louise Michel, et non l'accumulation de connaissances à visée exhaustive. Les analogies permet- tent de relier entre eux les différents champs du savoir, et sont donc conçues comme un véritable mode d'accès à la connaissance capable de guider l'huma- nité vers la « race », autrement dit vers l'émancipation. Aux sciences naturelles succèdent des récits de « mythes et légendes des temps anciens » ou encore une description de « grandes migrations historiques », l'ensemble relevant davantage de la vue générale que de la description détaillée. Cette conception se justifie bien sûr par la particularité du public visé, à savoir des enfants, qu'il s'agit de ne pas

122. IISHA, Louise Michel Papers, 568, « Encyclopédie enfantine », 1879–80.

123. Michel, *Lectures encyclopédiques par cycles attractifs* ; Michel, *Notions encyclopédiques par ordre attractif.*

124. IISHA, Louise Michel Papers, 568, « Encyclopédie enfantine », 1879–80, image numérisée n°10–11.

125. Cette analogie animalière du « troupeau humain » est un motif récurrent que l'on retrouve également dans l'œuvre littéraire de Louise Michel, à des fins de propagande anarchiste. Voir Broussais, « Diffu- ser l'anarchisme par la fiction ».

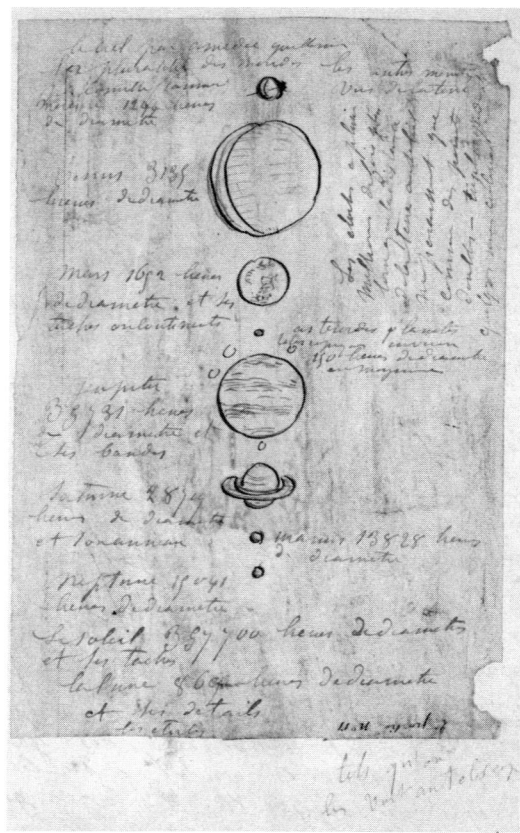

FIGURE 2 ISHA, Les principaux astres du Système solaire dessinés par Louise Michel, « tels qu'on les voit au télescope », Louise Michel Papers, 568, « Encyclopédie enfantine », 1879–80.

noyer sous un trop grand nombre de notions mais chez qui il faut plutôt susciter la curiosité la plus large possible. Louise Michel décrit son projet comme une « mappemonde »[126], non au sens littéral du terme (bien que quelques cartes y figurent) mais comme la métaphore de l'objet permettant non seulement d'avoir une vue globale sur le monde connu, mais également susceptible de susciter l'envie de le parcourir et de le découvrir par soi-même.

La méthode pédagogique défendue par Louise Michel dans les pages de l'« Encyclopédie enfantine » se caractérise notamment par une importante promotion de l'expérience sensible comme voie d'accès à la connaissance. C'est particulièrement le cas pour les parties du manuscrit consacrées à la géologie et à l'astronomie. Elle liste notamment une série d'observations possibles de différents reliefs et formations littorales témoignant du passé géologique de la France, ou décrit encore avec beaucoup de précision les différents astres composant le système solaire, en s'appliquant à les reproduire sous forme de dessins tels qu'ils apparaissent derrière l'oculaire d'un télescope[127]. La perspective de toucher un public n'ayant pas les moyens d'acquérir un instrument optique est ainsi prise en compte, en permettant aux petits apprenants d'acquérir malgré tout une représentation la plus proche possible de la réalité accessible par la pratique savante d'observation (fig. 2).

Cependant, ce choix de (re)présentation des différents astres du système solaire semble également être guidé par l'analogie comme méthode d'accès au

126. Louise Michel, *Lettre datée du 24 octobre 1878*, dossier « lettres inédites ».
127. IISHA, Louise Michel Papers, 568, « Encyclopédie enfantine », 1879–80, image numérisée n°1 à 4 et 72 à 79.

savoir : Louise Michel souligne en effet fréquemment les similitudes entre certaines formations géologiques de certaines planètes et celles observées sur Terre, de même qu'elle formule des hypothèses concernant les climats de Vénus ou de Mars, mais en les reliant à son expérience sensible personnelle :

> Mars dont les pôles sont couverts de glace a une lumière rougeâtre. Plus loin que nous du soleil, il doit, si les conditions sont les mêmes recevoir moins de chaleur que la Terre. Mais sur la Terre même, certaines circonstances modifient considérablement un climat. Ainsi la Calédonie où le sol devrait être brûlant est rafraîchie par les brises de mer. Encore bien plus d'un monde à l'autre doivent les mêmes circonstances existant les modifier diversement. D'un autre globe qui sait même si les habitants ont besoin des mêmes conditions que nous[128].

Le projet d'« Encyclopédie enfantine » révèle ainsi un projet politique d'enseignement condensant la conception particulière des savoirs et de leur transmission chez Louise Michel. En effet, les apprentissages s'appuient sur des éléments de connaissances validés et admis—puisque circulant dans les ouvrages—et sur le schème classique de leur accumulation—la bibliothèque constituant le point de départ autant que la condition de possibilité de l'enseignement. Le projet d'émancipation se fonde donc, ici, sur des schèmes pédagogiques classiques. Toutefois, leur dimension politique est manifeste d'abord dans les réseaux mobilisés pour constituer le fonds culturel d'apprentissage qu'est la bibliothèque, ensuite dans l'usage réitéré de l'analogie comme modalité argumentative capable de rendre sensible l'appréhension savante du monde.

Conclusion

Louise Michel s'est donc obstinée, dans sa visée savante, à pénétrer des espaces sociaux connectés mais fort différents : le bagne (à la fois lieu de collecte botanique, espace de proximité avec les autochtones et siège de son enseignement), les institutions savantes (la Société d'acclimatation) et les milieux militants (qui lui fournissent notamment des ressources livresques). Ce qui frappe, dans la recomposition des différentes traversées socio-épistémiques de Louise Michel, c'est sa capacité à s'ajuster *partiellement* aux attentes de chacun d'eux. Elle est ainsi capable de construire un programme cohérent de collection et d'expérimentation botanique—même si ses essais de vaccination végétale sont clairement en-dehors des canons attendus de la science de son temps. Elle peut passer par ses relais militants pour faire circuler des échantillons, diffuser ses savoirs ethnographiques ou faire reconnaître ses productions pédagogiques—même si ses buts

---

128. IISHA, Louise Michel Papers, 568, « Encyclopédie enfantine », 1879–80, image numérisée n°75.

ne sont pas toujours atteints. Enfin, elle éprouve les logiques pénitentiaires, jusque dans leur plasticité en tentant d'élargir ses terrains d'enquête et en faisant du bagne un espace possible d'apprentissage—sans toujours parvenir à ses fins. Deux fils épistémiques continus sous-tendent l'attitude savante de Louise Michel : d'abord son approche sensible de la connaissance, sa manière concrète d'aborder les savoirs et de les transmettre ; ensuite, la matrice analogique de son système d'explication qui va jusqu'à lui faire mettre en correspondance les comportements zoologiques et les politiques humaines. Ces deux traits caractéristiques expliquent, au moins en partie, pourquoi Louise Michel parvient à traverser et à mobiliser des espaces socio-épistémiques si différents. D'une part, les savoirs sensibles n'ont pas encore disparu à la fin du XIXe siècle, même s'ils sont mis à mal ; ils restent encore légitimes pour bon nombre d'approches savantes du monde, en particulier dans les milieux amateurs et militants. Ensuite, l'analogie n'a pas encore totalement été récusée comme mode d'intellection, elle subsiste encore dans certains domaines de la physique par exemple[129].

Louise Michel ne correspond donc ni au modèle d'une chercheuse académique, ni à celui d'une militante orientant tous ses efforts d'apprentissage et de connaissance vers l'action politique—même si elle peut occasionnellement le faire. Elle reconnaît la légitimité des sciences académiques mais ne se sent pas contrainte par les règles qui en restreignent l'accès et peut s'en affranchir sans crainte. C'est cette capacité à *partiellement* investir des espaces socio-épistémiques distincts qui donne à Louise Michel la possibilité de circuler entre eux, sans être attachée strictement à aucun d'eux. Mais si Louise Michel n'a jamais pu *totalement* investir les espaces académiques, c'est d'abord parce que l'université est encore interdite aux femmes et que le patriarcat a édifié la science en régime purement masculin.

Par cette liberté épistémique qu'elle revendique envers et (parfois) contre tous, la « Minerve populaire » met ainsi en pratique un engagement anarchiste universaliste, valorisant constamment l'émancipation individuelle et collective, l'attention profondément empathique à tous les êtres, ainsi que l'égalité universelle des capacités.

VOLNY FAGES est maître de conférences en épistémologie et histoire des sciences à l'Ecole Normale Supérieure Paris-Saclay. Il est l'auteur de *Savantes nébuleuses* (2018).

JÉRÔME LAMY est historien et sociologue des sciences, directeur de recherche au CNRS (CESSP, EHESS). Il a publié, avec Jean-François Bert, *Voir les savoirs : Lieux, objets et gestes de la science* (2021).

129. Galison et Assmus, « Artificial Clouds, Real Particles ».

FLORIAN MATHIEU est docteur en histoire des sciences, chercheur associé à l'Université Paris-Saclay. Il est l'auteur d'une thèse intitulée « Usages politiques et populaires du savoir astronomique : Entre science et utopies révolutionnaires (France, 1871–1939) » (2022).

Références

Baldwin, Melinda. *Making "Nature" : The History of a Scientific Journal*. University of Chicago Press, 2015.

Baylac, Marie-Hélène. *Louise Michel*. Perrin, 2024.

Bensa, Alban. *En pays kanak : Ethnologie, linguistique, archéologie, histoire de la Nouvelle-Calédonie*. Editions de la Maison des Sciences de l'Homme, 2000.

Bensaude-Vincent, Bernadette, et Anne Rasmussen, dirs. *La science populaire dans la presse et l'édition (XIXe–XXe siècle)*. CNRS Editions, 1997.

Bertrand, Romain. *Le détail du monde : L'art perdu de la description de la nature*. Seuil, 2019.

Blanqui, Auguste. *Instructions pour une prise d'armes : L'éternité par les astres ; Hypothèse astronomique, et autres textes*. Editions de la Tête de Feuille, 1972.

Broussais, Romain. « Diffuser l'anarchisme par la fiction : *La chasse aux loups* (première partie) de Louise Michel ». *Revue droit et littérature* 7, n°1 (2023) : 161–77.

Bullard, Alice. *Exile to Paradise : Savagery and Civilization in Paris and the South Pacific, 1790–1900*. Stanford University Press, 2000.

Cahan, David. *From Natural Philosophy to the Sciences : Writing the History of Nineteenth-Century Science*. University of Chicago Press, 2003.

Caton, Joannès. *Journal d'un déporté de la Commune à l'île des Pins*. France-Empire, 1986.

Cerf, Marcel. *Le mousquetaire de la plume : La vie d'un grand critique dramatique, Henry Bauër, fils naturel d'Alexandre Dumas, 1851–1915*. Cerf, 1975.

Chapman, Allan. *The Victorian Amateur Astronomer : Independent Astronomical Research in Britain, 1820–1920*. Wiley, 1998.

Charvolin, Florian, Aurélie Dumain, et Jacques Roux, dirs. *Passions cognitives : L'objectivité à l'épreuve du sensible*. Editions des Archives Contemporaines, 2013.

Clair, Sylvie. « " User le soleil avec la pierre ponce " ou la vie des déportés communards en Nouvelle-Calédonie d'après leur correspondance (1872–1880) ». *Histoire de la justice*, n°5 (1992) : 117–51.

Conner, Clifford D. *Histoire populaire des sciences*. Editions L'Echappée, 2011.

Dauphiné, Joël. *La déportation de Louise Michel*. Les Indes Savantes, 2006.

Davagnier, Jean-Baptiste. *Le bon vieux temps*. Perreau, 1877.

Davagnier, Jean-Baptiste. *Le carnaval de nos jours*. Panvert, 1879.

Davagnier, Jean-Baptiste. *Le réveil du peuple; ou, Le dernier des rois*. Guillois et Cie, 1848.

De Martonne, Ed. *Tables générales de « La Géographie », bulletin de la « Société de géographie », 8e, 9e, 10e, et 11e série, 1900–1939*. Paris, 1945.

Eichner, Carolyn J. *Feminism's Empire*. Cornell University Press, 2022.

Eichner, Carolyn J. « Language of Imperialism, Language of Liberation : Louise Michel and the Kanak-French Colonial Encounter ». *Feminist Studies* 45, n°2–3 (2019) : 377–408.

Fages, Volny. *Savantes nébuleuses : L'origine du monde entre marginalité et autorité scientifique (1860–1920)*. Editions de l'EHESS, 2018.

Fierro, Alfred. *Inventaire des manuscrits de la Société de géographie*. Bibliothèque nationale de France, 1984.

Frosini, Fabio. « Le travail caché du prisonnier entre " littérature " et " politique " : Quelques réflexions sur les " sources " des *Cahiers de prison* d'Antonio Gramsci ». *Laboratoire italien*, n°18 (2016). https://journals.openedition.org/laboratoireitalien/1064.

Galison, Peter, et Alexi Assmus. « Artificial Clouds, Real Particles ». Dans *The Uses of Experiment*, dirigé par David Gooding, Trevor Pinch, et Simon Schaffer, 225–74. Cambridge University Press, 1989.

Gieryn, Thomas. *Cultural Boundaries of Science : Credibility on the Line*. University of Chicago Press, 1999.

Ginzburg, Carlo. *Le fromage et les vers : L'univers d'un meunier du XVIe siècle*. Aubier, 1980.

Guillemain, Hervé, et Nathalie Richard. « Introduction : Towards a Contemporary Historiography of Amateurs in Science (Eighteenth–Twentieth Century) ». *Gesnerus* 72, n°2 (2016) : 201–37.

Hart, Kathleen. « Oral Culture and Anticolonialism in Louise Michel's *Mémoires* (1886) and *Légendes et chants de gestes canaques* (1885) ». *Nineteenth-Century French Studies* 30, n°1–2 (2001–2) : 107–20.

Judde de Larivière, Claire. *L'ordinaire des savoirs : Une histoire pragmatique de la société vénitienne (XVe–XVIe siècles)*. Paris, 2023.

Lamy, Jérôme. « La béance : Notes, carnets et brouillons de Germaine Tillion ». *Ethnologie française*, n°49 (2019) : 597–616.

Leenhardt, Maurice. *Langues et dialectes de l'Austro-Mélanésie*. Institut d'Ethnologie, 1946.

L'égalité. « Le suffrage universel : Conférence de Mlle Louise Michel, à la salle du boulevard des Capucines ». N°71 (1889) : 3.

Lemosof, Paul. *Table des matières, séries V–VII, Bulletin de la Société de géographie, 1861–1899*. Société de Géographie, 1905.

Licoppe, Christian. *La formation de la pratique scientifique : Le discours de l'expérience en France et en Angleterre (1630–1820)*. La Découverte, 1996.

Marmo Mullaney, Marie. « Sexual Politics in the Career and Legend of Louise Michel ». *Signs*, n°2 (1990) : 300–322.

Méhaud, Catherine. *Manuscrit de la Société de géographie concernant l'Asie et l'Océanie*. Bibliothèque nationale de France, 1979.

Michel, Louise. *Je vous écris de ma nuit : Correspondance générale de Louise Michel, 1850–1904*. Dirigé par Xavière Gauthier. Editions de Paris, 1999.

Michel, Louise. *Lectures encyclopédiques par cycles attractifs*. Librairie d'éducation laïque, 1888.

Michel, Louise. *Légendes et chansons de gestes canaques (1875) suivi de Légendes et chants de gestes canaques (1885) et de Civilisation*. Textes établis et présentés par François Bogliolo. Presses Universitaires de Lyon, 2006.

Michel, Louise. *Mémoires de Louise Michel écrites par elle-même*. Tome 1. F. Roy, 1886.

Michel, Louise. *Notions encyclopédiques par ordre attractif*. P. Buchillot, 1894.

Narayana, Valérie. « Objets trouvés d'une réprouvée : Les objets d'écriture de Louise Michel en Nouvelle-Calédonie ». *Nouvelle revue synergies Canada*, n°13 (2020). https://www.erudit.org /en/journals/nrsc/2020-n13-nrsc06130/1078430ar/.

Panziera, Sophie. « L'exaltation de la veille politique : Vigilance et veille citoyenne de la Révolution française à la Commune de Paris ». Dans « Le sommeil au XIXe siècle : Normes et imaginaires du dormir (années 1770–1914) », 51–152. Thèse de doctorat, Université Paris 1, 2023.

Rancière, Jacques. *La nuit des prolétaires*. Paris, 1981.

Roy, F. « Préface de l'éditeur ». Dans tome 1 de *Mémoires de Louise Michel écrites par elle-même*, i–viii. Paris, 1886.

Rustenholz, Alain. *Paris ouvrier, des sublimes aux camarades*. Parigramme, 2003.

Salas, Denis. « Louise Michel déportée politique au bagne de Nouvelle-Calédonie (1874–1880) ». *Histoire de la justice*, n°15 (2002) : 239–49.

Secord, Ann. « Science in the Pub : Artisan Botanists in Early Nineteenth-Century Lancashire ». *History of Science* 32, n°3 (1997) : 269–315.

Soula, Virginie. *Histoire littéraire de la Nouvelle-Calédonie (1853–2005)*. Karthala, 2014.

Verhaeghe, Sidonie. « Louise Michel, féministe : Analyse d'une opération de qualification politique aux débuts de la IIIe République ». *Le temps des médias*, n°29 (2017) : 18–32.

Les toilettes pour dames s'emparent du macadam

Une étude architecturale et ingénieuriale des chalets de nécessité parisiens (1872–1900)

LOUISE THIROUX

PRÉCIS   Cet article retrace l'histoire d'une typologie méconnue du mobilier urbain parisien : les chalets de nécessité. Se déclinant sous deux principales formes, ces lieux d'aisance fleurissent dans la capitale entre 1872 et 1900. L'histoire socioculturelle des techniques et de l'architecture contribue à souligner les différentes intrications que le mobilier de rue entretient avec les discours hygiénistes de l'époque, forgeant de ce fait une place toute particulière pour ce que nous appellerons le *corps pissant* féminin en cette fin de siècle.

MOTS CLÉS   toilettes publiques, Paris, histoire des techniques, histoire urbaine, histoire des femmes

Longtemps restée sourde aux besoins des vessies féminines, la municipalité parisienne décide, en 1872, d'installer à titre d'essai quatre petits kiosques ou chalets de nécessité à la suite du rapport établi par Théodore-Jacques Bonvalet, membre du parti radical au conseil municipal[1]. Les femmes ont donc dû attendre trente-sept années de plus que les hommes avant de disposer de lieux d'aisance, à croire qu'entre la Monarchie de Juillet et le Second Empire, « l'intérêt porté au sexe féminin ne comportait aucune part vésicale »[2]. Entre 1872 et 1900, cette typologie architecturale se déploie dans deux principaux modèles installés sur la chaussée. Nous en comptons 103 à la fin de la période, contre 1703 urinoirs[3]. Le choix d'offrir des chalets aux femmes dès les années 1870 fait de Paris une ville pionnière en Europe. Les Londoniennes devront, par exemple, attendre jusque dans les années 1890 pour bénéficier de toilettes exclusivement souterraines et

---

1. *Le XIXᵉ siècle, journal quotidien, politique et littéraire*, juil. 1872.
2. Maillard, *Les vespasiennes de Paris, ou les Précieux édicules*, 45.
3. *Le XIXᵉ siècle, journal quotidien, politique et littéraire*, août 1898.

d'une grande discrétion, baptisées *lavatories*. Si les options diffèrent, elles soulignent cependant la nécessité de doter les villes d'infrastructures sanitaires afin de palier leur manque dans l'habitat privé et leur faible nombre dans les lieux de travail, les magasins ou encore les théâtres. L'étude architecturale et ingénieuriale des chalets de nécessité révèle à la fois les tendances urbanophobes à l'œuvre dans le dernier quart du XIXᵉ siècle, tout en documentant la progressive bascule de l'hygiène publique en hygiène sociale. En outre, elle met en lumière la manière dont est conçue et perçue la place du *corps pissant* féminin dans Paris à partir des années 1870.

Il s'agira ici de montrer comment cette typologie du mobilier urbain tente d'incarner et de diffuser des logiques d'hygiène, de pudeur et de confort, marquant un cran supplémentaire dans la volonté d'octroyer une place de plus en plus réduite et discrète au *corps pissant* dans la capitale. Afin d'appréhender au mieux ces différents aspects, ce travail suit un plan thématico-typologique. Les chalets de type I, commodités ouvertes tant aux hommes qu'aux femmes, seront tout d'abord envisagés comme des lieux dont la visée est de restaurer un ordre social et moral par des logiques de contrôle tant architecturales qu'ingénieurales. Puis, nous nous intéresserons à la manière dont les chalets de type II, payants et strictement réservés aux dames, contribuent à définir la place du *corps pissant* féminin en associant le confort et la discrétion aux préceptes hygiénistes en vogue à cette époque.

Des lieux anti-urbains bien urbains : les chalets de nécessité type I

Inauguré place de la Bourse le 26 septembre 1873 (fig. 1), le tout premier chalet de nécessité mixte implanté sur les trottoirs parisiens suscite l'enthousiasme de la presse parisienne. *La petite presse : Journal quotidien* y consacre un article entier dans son numéro du 27 septembre 1873, se concluant sur le souhait « que le type du chalet de la Bourse sera reproduit à un nombre suffisant d'exemplaires »[4]. Hélas, selon le *Rapport municipal de Paris*, nous ne dénombrons que quatre chalets de nécessité mixtes en 1879[5], soit deux de moins que ce qui était prévu en 1872. Au lieu de se situer boulevard Bonne-Nouvelle, place de la Madeleine, place du Théâtre-Français, place du Château d'Eau, place de l'Etoile et place de la Bourse[6], nous les retrouvons au Marché aux fleurs de la Cité, place de la Bourse, place de la Madeleine et aux Champs-Elysées[7]. Loin d'être neutre, la géographie initiale du projet atteste de la volonté d'offrir des lieux d'aisance dans des quartiers populeux et dans ceux où les femmes travaillent et se divertissent

---

4. *La petite presse*, sept. 1873.
5. *Rapport du conseil municipal de Paris*, janv. 1879.
6. *La petite presse*, janv. 1873.
7. *Rapport du conseil municipal de Paris*, janv. 1879.

FIGURE 1 Georges Chevalier (1882–1967). Paris (IIe arr.), France. La place de la Bourse, vue de la rue Notre-Dame-des-Victoires, 1914. Cote : Albert Kahn, A7557.

majoritairement. L'histoire des chalets de nécessité permet en outre de redonner aux femmes la visibilité et la mobilité qu'elles avaient dans l'espace public dans les quartiers de la confection (boulevard Bonne-Nouvelle), les quartiers des théâtres ou encore des grands magasins (place du Théâtre-Français, place de la Bourse, etc.), déclinant ainsi au féminin les notions de travailleur, de flâneur et de pisseur.

Si ces chalets peinent à prendre place dans la ville, c'est en partie en raison de leur incongruité; la cohabitation des sexes n'allant pas de soi à cette époque. Le conseiller municipal Jacques Weber est plus qu'éloquent à ce sujet : « Dans les couloirs toujours trop étroits des chalets, les deux sexes se coudoient, non sans quelques froissements et surtout non sans toujours choquer le sentiment très naturel et très légitime de la pudeur féminine »[8]. Il s'agit alors de saisir comment les architectes, les ingénieurs, la municipalité et les usagers tentent de dépasser cette initiale étrangeté en conférant à ces lieux une forme d'urbanité. Pour ce faire, la forme anti-urbaine et composite des chalets de type I sera analysée. Puis, nous appréhenderons la façon dont ces commodités se parent des différents atours de la modernité afin de restaurer leur image. Enfin, nous constaterons que ces lieux d'aisance demeurent malgré tout impopulaires et suspects.

Le chalet : Une forme anti-urbaine et composite

Remises au goût du jour par des historiens tels que Manuel Charpy, les études portant sur les chalets en ville se sont majoritairement concentrées sur la question des « constructions individuelles dans les quartiers périphériques »[9] et non sur la typologie de ce mobilier urbain « monumental ». Il apparaît pourtant indéniable qu'elle est tributaire du goût pour l'architecture alpine triomphant dans les années 1870, comme l'a souligné Michel Vernes[10]. Triomphe qui peut s'observer à la lecture des différents numéros de la *Revue générale de l'architecture et des travaux publics* de cette décennie, mettant en avant tant la dimension « pittoresque »[11] de l'architecture helvétique que l'intérêt croissant pour les chalets[12]. Avant de se pencher sur ses traductions formelles, il convient d'interroger la signification de l'incursion du chalet dans le paysage urbain. En effet, cette mode ne saurait être comprise sans le recours au concept d'anti-urbanité, axe central de la réflexion de François Loyer sur ce type d'architecture[13]. La pensée anti-urbaine traduit une hostilité à la ville, fondée sur le postulat qu'elle serait néfaste pour la société, par opposition à la montagne. Par ailleurs, cette notion s'articule aux préceptes hygiénistes, ainsi résumés par Albert Lévy : « Pour l'idéologie urbanistique hygiéniste naissante, consolidée ensuite par les travaux Pasteur (1822–1895), il s'agissait de " combattre " à la fois le milieu physique, insalubre, hostile, malsain, porteur de germes pathogènes, et le milieu social, les classes populaires, jugées dangereuses, causes de troubles politiques et sociaux,

---

8. *Rapport du conseil municipal de Paris*, févr. 1901.
9. Loyer, *Paris XIXᵉ siècle*, 340.
10. Vernes, « Le chalet infidèle ».
11. Daly, *Revue générale de l'architecture et des travaux publics*, t. 35.
12. Daly, *Revue générale de l'architecture et des travaux publics*, t. 34.
13. Loyer, *Paris XIXᵉ siècle*, 343.

favorisant également la contagion »[14]. Ce concept peut aussi s'appliquer aux chalets de type I dont la fonction et la mixité sont contraires à la moralité. Cette forme, appliquée aux commodités publiques, revêt deux objectifs : signaler et pallier l'incongruité des lieux. Le « mythe des Alpes »[15] nourrit alors une fiction ingénieuriale visant à corriger la ville. Elle se manifeste dans les toilettes publiques par sa « toiture de chalet suisse »[16], symbole du « rêve improbable de réconciliation de la ville et de la campagne, de l'individu et de la société »[17]. En outre, l'activation d'un imaginaire helvétique participe de la bonne marche du lieu, en rassurant l'usagère, en lui offrant un abri, « une solitude heureuse que la ville lui refuse et qui est l'occasion attendue de jouir librement de lui-même »[18]. Néanmoins, l'architecture de chalet ne peut se saisir indépendamment de son environnement, tant elle tend « à faire paysage »[19]. Ce n'est donc pas par hasard si le *Cahier des charges de la Concession de cent chalets de nécessité*[20], établi par le Service des concessions sur la voie publique en 1879, préconise l'installation de ces édifices « sur les grands terre-pleins plantés, dans les grands squares et sur les voies plantées dont les trottoirs ont une largeur de 8 mètres, au minimum ». Ainsi, ces lieux accompagnent des espaces de végétation en ville, dont le meilleur représentant est le chalet de type I des Champs-Elysées, immortalisé par Charles Marville (fig. 2). Nous saisissons alors comment ce genre d'édicule se positionne au sein d'une nature encadrée et contrôlée. Ce simulacre de nature, inséré dans la fiction ingénieuriale, sert à son tour la rhétorique urbanophobe en confortant les prétendues vertus salvatrices de la montagne. Vertus dont nul n'est dupe, surtout pas la presse qui voit dans cette stratégie plus une volonté de dissimulation qu'une opération salutaire pour la salubrité publique. *La presse*, dans son édition du 12 août 1892, écrit à ce sujet : « Ne dissimule-t-on pas les chalets de nécessité sous des bosquets de fleurs ? »[21]. Rien ne semble donc pouvoir racheter l'inconvenance de ces chalets mixtes. De surcroît, les chalets de type I sont de style très composite et hétérogène. L'article 2 du *Cahier des charges de la Concession de cent chalets de nécessité* va en ce sens : « Les chalets pourront présenter des dispositions variées, soit en plan, soit en élévation ». Cette propension au mélange est en adéquation avec l'architecture de chalet tant sa « migration a entraîné l'hybridation progressive de [son] architecture. [...] En s'acclimatant à la ville, elle a assimilé dans le désordre tous les styles appréciés des citadins »[22]. Cette

14. Lévy, « L'intégration de la santé dans l'éco-urbanisme », 113.
15. Salomon Cavin, « La ville mal-aimée ».
16. *La France*, janv. 1881.
17. Vernes, « Le chalet infidèle », 136.
18. Vernes, « Le chalet infidèle », 113.
19. Vernes, « Le chalet infidèle », 122.
20. Archives de Paris, VONC 16.
21. *La presse*, août 1892.
22. Vernes, « Le chalet infidèle », 117.

FIGURE 2 Charles Marville (1813–79). Cabinet Water-Closet Dorion (Champs-Elysées), v. 1865, Paris. Cote : Musée Carnavalet, PH2744.

hybridation est lisible à travers la question de l'ornement architectural. Il n'est pas anodin que le recours à l'ornementation soit systématique dans les toilettes à destination féminine et sporadique dans celles à destination masculine. L'interprétation qu'en donne Antoine Picon est des plus éclairantes : « D'après les stéréotypes bourgeois, là encore, l'ornement se révèle également plus féminin que masculin, car selon l'esprit positiviste, la constitution de la femme semble moins encline à la raison que son pendant masculin »[23]. Cette utilisation s'inscrit donc dans une tentative de séduction du public féminin auquel s'adressent ces lieux mixtes. En outre, l'ornement, qui vit son âge d'or durant les trois premiers quarts du XIXᵉ siècle, se complaît dans des formes orientalisantes, dont les Parisiennes sont friandes. Bien que nous n'ayons pas conservé de traces matérielles et visuelles de ces édicules mâtinés d'orientalisme, la presse d'époque nous en fournit des descriptions. *Le corsaire : Journal quotidien* évoque l' « élégante plate-forme chinoise, ornée aux quatre coins de quatre urnes »[24] des chalets en devenir, tandis que *Le charivari* fait état du « système oriental »[25] de l'édicule installé près du

23. Picon, *L'ornement architectural*, 112.
24. *Le corsaire*, août 1872.
25. *Le charivari*, nov. 1881.

théâtre de l'Ambigu-Comique. Si l'emploi de motifs orientalisants participe à l'élégance des lieux, il renforce aussi son étrangeté. Daniel-Henri Pageaux et Christine Peltre perçoivent dans l'orientalisme l'occasion de présenter un « Occident inversé, confiné dans une irréductible altérité : non la raison, mais la passion, le merveilleux, la cruauté ; non le progrès ou la modernité, mais le temps arrêté, le primitif »[26]. Cette inversion coïncide avec celle des valeurs morales que présupposent des commodités où les deux sexes peuvent assouvir des besoins perçus de manière croissante comme privés à partir des années 1870. Si les allures bigarrées des chalets semblaient séduire jusque dans les années 1880, la situation est toute autre à la fin du siècle. Certains journaux appellent à leur démolition, non pour des raisons morales, mais esthétiques : « Il faut y mettre bon ordre, si l'on n'oblige pas M. Picard à tout démolir à bref délai, nous sommes exposés à conserver, parmi tant de chalets de nécessité qui déshonorent nos promenades, cette galerie vraiment trop hétéroclite de chalets sans nécessité »[27]. Cette condamnation est à mettre en perspective avec celle de l'ornement, perçu comme contraire à la modernité, à la fin du siècle[28]. Une modernité que ces édifices s'efforcent pourtant de refléter afin de restaurer leur urbanité.

Une urbanité recomposée : La matérialité des chalets de type I

Bien que leur forme soit anti-urbaine, l'aménagement et le mode de fonctionnement des chalets de type I ne sont en rien conformes avec ce que nous pourrions attendre de semblables constructions, tant ils suivent les logiques d'urbanité propres au dernier tiers du XIXe siècle. Par *urbanité*, nous entendons ce qui a trait au raffinement des manières, mais aussi ce qui concerne le caractère urbain d'un espace, à savoir sa mixité sociale et fonctionnelle, indissociable des exigences hygiénistes et morales qui imposent leurs vues régulatrices à la capitale. Nous verrons alors comment l'hygiénisme dicte tant la structure que le fonctionnement des chalets, tout en laissant la place à une certaine pensée du luxe propice à la mise en scène de la décence entre les sexes.

La pensée hygiéniste se retrouve à l'œuvre tant à l'extérieur qu'à l'intérieur des édicules. C'est pourquoi chaque chalet donne à voir sur sa toiture « une ou deux lanternes d'aération »[29] ou encore « un couronnement de fer ouvré pour le passage de l'air »[30]. Nous comprenons alors comment s'inscrit architecturalement la question de l'aération, pivot de la pensée aériste du XVIIIe siècle[31],

---

26. Pageaux et Peltre, « orientalisme, art et littérature », Encyclopædia Universalis [en ligne], consulté le 15 avril 2023. https://www.universalis.fr/encyclopedie/orientalisme-art-et-litterature/.

27. *La presse*, déc. 1900.

28. Picon, *L'ornement architectural*.

29. *La France*, oct. 1879.

30. *Le XIXe siècle*, juin 1873.

31. Delamare, *Traité de la police*.

réinvestie par les traités hygiénistes au XIXᵉ siècle[32]. Dès lors, en rendant visible de l'extérieur son système d'aération, le chalet mixte met en scène sa propre salubrité afin de rassurer ses potentiels clients. Cependant, les préceptes hygiénistes s'appliquent aussi au fonctionnement interne des chalets via le système d'évacuation des matières. L'article 3 du *Cahier des charges de la Concession de cent chalets de nécessité* en date de 1879 est dédié à cette question. Il stipule que « les appareils de vidange seront des tinettes filtres et les eaux vannes seront conduites à l'égout par un tuyau en fonte ou en poterie résistante, de 0m.20 au moins de diamètre, à fermeture hydraulique. Les fosses mobiles seront seules tolérées dans les emplacements où l'évacuation à l'égout serait impossible ». Cet article résume l'ensemble des préconisations et des acquis de l'haussmannisation. Tout d'abord, l'embranchement à l'égout est pensé selon le modèle haussmannien de l'exclusion des excréments[33], comme le suggère la présence de tinettes filtrantes. Ces appareils diviseurs, brevetés depuis les années 1850,[34] permettent de séparer les matières solides des matières liquides afin de mener ces dernières seulement à l'égout. Imparfait, ce système constitue, selon *Le génie civil : Revue générale des industries françaises et étrangères* du 15 septembre 1883, un moindre mal pour la salubrité des latrines : « Ces inconvénients existent en effet, mais on les a considérablement exagérés. On commet d'ailleurs une grande erreur en les attribuant au système lui-même »[35]. Rassurantes, les tinettes permettent à l'égout et par extension, aux chalets de type I de rejoindre « ce monde métonymique que l'imaginaire de l'ordre superpose au monde réel pour rendre compte du progrès »[36]. Néanmoins, les chalets de type I sont rapidement adaptés au tout-à-l'égout, préconisé par le rapport général du 18 juillet 1883 de la Commission technique de l'assainissement de Paris[37]. La liste des emplacements, fournie par la Société anonyme des chalets de nécessité en novembre 1885, indique quels édicules sont déjà desservis par l'écoulement direct ainsi que tous ceux qui pourront bientôt l'être. Sur les soixante-dix chalets mixtes existant à cette date, cinquante-quatre sont déjà reliés au tout-à-l'égout et les autres le seront à la fin des années 1880, soit quatre années avant que le préfet Poubelle ne rende ce système obligatoire. Ainsi, en se pliant à la pensée technique hygiéniste, les chalets de type I s'inscrivent dans une logique d'urbanité sous-tendue par la modernité de leur infrastructure. Une urbanité d'ailleurs renforcée par la matérialité des lieux. En effet, la matérialité de ces édifices emprunte à la fois d'une

32. Becquerel, *Traité élémentaire d'hygiène privée et publique.*
33. Haussmann, *Mémoires*, t. 3.
34. INPI, 1BB23156, Bonamour, Système d'appareil séparateur et filtre applicable aux fosses d'aisance, 1855.
35. *Le génie civil*, sept. 1883.
36. Laroulandie, « Les égouts de Paris au XIXᵉ siècle », 130.
37. *Assainissement de Paris : Résolutions votées par la commission technique de l'assainissement de Paris et résumé des travaux de la commission*, Paris, 1883.

pensée du luxe, mais aussi de la modernité, entendue comme « la tradition du nouveau »[38]. La presse est plus que disserte à ce sujet. *La petite presse*, dans son édition du 27 septembre 1873, mentionne le « grand luxe »[39] avec lequel les chalets de type I sont installés. Cela s'explique par les matériaux utilisés dans les chalets, surtout le marbre. Ce matériau noble orne les cloisons extérieures[40] de ces lieux d'aisance, dont l'exemple le plus fameux est l'édicule situé boulevard de la Madeleine. Nous percevons alors combien la satisfaction du goût agit comme un prérequis à la satisfaction des besoins. Parmi ces prérequis, nous retrouvons également l'exigence d'une forme de modernité, qui transparaît par l'utilisation du fer et du verre. Idéal pour les « constructions qui visent à des buts transitoires »[41], le fer en tant que « symbole esthétique de la grande industrie naissante qui sut au siècle dernier incorporer ses produits dans les constructions les plus banales en leur donnant un sens nouveau dépassant celui de leur usage courant »[42] matérialise l'idée de modernité. Au-delà de son aspect symbolique et de son bas coût, le fer présente aussi un certain nombre de propriétés constructives intéressantes sur lesquelles revient Sigfried Giedion : « Le fer est capable d'extension ou de compression. Il offre une résistance à la tension et à la pression, et en conséquence, à la courbure. Son privilège : pouvoir satisfaire à un maximum d'exigences en mobilisant des dimensions minimales »[43]. Toutefois, sa modernité constructive ne saurait être complète sans l'association qu'elle entretient avec un autre matériau : le verre. Si l'intérêt d'une telle combinaison est pressenti dès 1849 par la *Revue générale de l'architecture et des travaux publics*[44], elle se concrétise pleinement dans les chalets de type I. L'article 2 du *Cahier des charges de la Concession de cent chalets de nécessité* mentionne des « panneaux vitrés translucides, mais non transparents ». Cette évocation révèle toute l'ambiguïté de l'ambiance qui règne tant au seuil qu'au sein même des édicules. Relevée par Jean Baudrillard dans *Le système des objets*, l'ambiguïté du verre est « d'être à la fois proximité et distance, intimité et refus de l'intimité, communication et non communication »[45]. Plus encore, il souligne aussi le potentiel moral, hygiénique et prophylactique qui fait du verre « le matériau de l'avenir »[46]. C'est dans leur utilisation de ce matériau que ces édifices se drapent dans le simulacre d'une modernité salvatrice, gageant tant de leur salubrité que de leur moralité. La

38. Rosenberg, *La tradition du nouveau*.
39. *La petite presse*, sept. 1873.
40. *La petite presse*, janv. 1873.
41. Benjamin, *Paris, capitale du XIXᵉ siècle*, 7.
42. Lemoine, *L'architecture du fer*, 278.
43. Giedion, *Construire en France*, 17.
44. Daly, *Revue générale de l'architecture et des travaux publics*, t. 8.
45. Baudrillard, *Le système des objets*, 58.
46. Baudrillard, *Le système des objets*, 58.

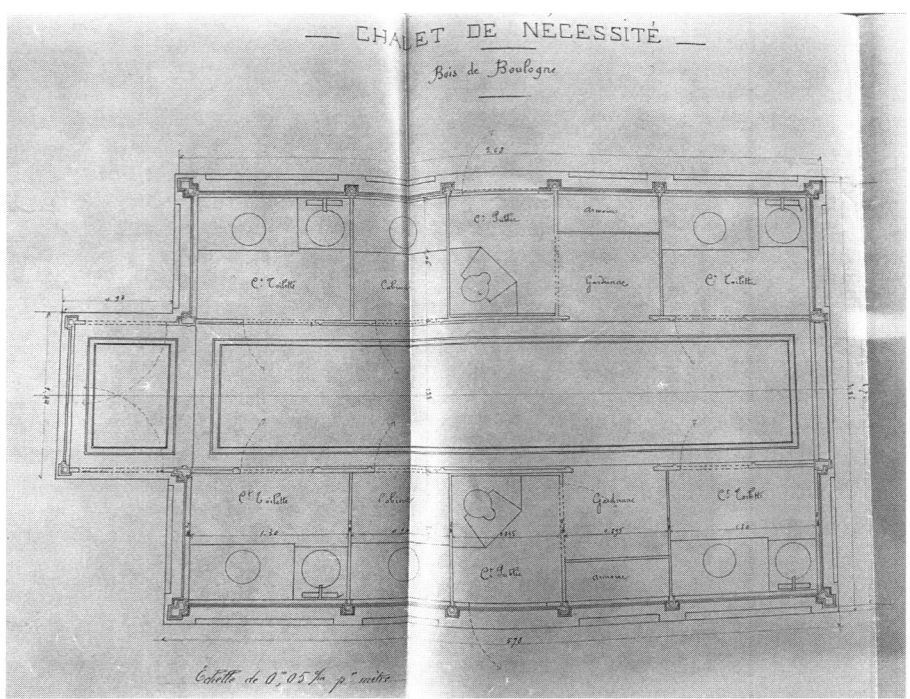

FIGURE 3 Anonyme. Plan chalet de nécessité de type I, v. 1875, Paris. Cote : Archives de Paris, V03 420.

question de la mise en scène de la moralité des lieux est essentielle à la bonne marche des chalets de type I. Bien qu'à destination des hommes et des femmes, la disposition des lieux tente de limiter au maximum les interactions entre ces derniers. Cette volonté de séparation s'affiche d'ailleurs dès l'entrée des édicules comme le souligne l'article de *La France : Politique, scientifique et littéraire* du 23 octobre 1879 : « Les premiers auront une entrée différente pour chaque sexe. Les seconds n'auront qu'une seule entrée »[47]. En outre, ce cloisonnement s'observe aussi à l'intérieur des latrines, comme le stipule le numéro du 3 janvier 1881 de *La France : Politique, scientifique et littéraire :* « La superficie qu'il occupera est de vingt mètres environ (cinq mètres sur quatre), dix stalles ou cabines y seront aménagées, à raison de cinq pour chaque sexe »[48]. Plus encore, l'étude des plans d'édification met en évidence l'existence d'un ou de plusieurs couloirs à l'intérieur des lieux afin d'éviter d'éventuels « frottements » entre les sexes (fig. 3). Ils ont d'ailleurs une double fonction : ils contribuent d'une part à rassurer et à

47. *La France*, oct. 1879.
48. *La France*, janv. 1881.

conforter les femmes dans leur pudeur tout en évitant d'autre part des rapprochements incongrus et immoraux dans ces lieux de passage. Qu'il s'agisse de leur décor, de leur système technique ou de leur manière de séparer les sexes, tout est mis en œuvre afin de restaurer l'urbanité des lieux. Pourtant, malgré toutes ces précautions, ces lieux ne parviennent pas à échapper à leur réputation sulfureuse et demeurent sous surveillance.

Des chalets étroitement surveillés

Bien que répondant à de nombreux critères d'urbanité, les chalets de type I ne semblent parvenir à se défaire de leur parfum de scandale corrélé aux différentes activités illicites qui y ont cours. Dès lors, il conviendra de revenir sur ces pratiques interlopes avant d'analyser comment les ingénieurs et la municipalité œuvrent afin d'enrayer ces dernières par un système de surveillance reposant sur la figure de la gardienne et de sa loge, mais aussi sur l'emploi de l'éclairage au gaz dans les édicules.

Les chalets de type I sont à bien des égards le théâtre de toutes les pratiques que le XIXᵉ siècle fustige : prostitution, homosexualité, vol, meurtre et suicide. Bien que conçus avec pour objectif de maintenir l'ordre, ces lieux constituent le siège de certains débordements, allant du cocasse au sordide. En premier lieu, les chalets n'échappent pas au spectre tenace de la prostitution. Ils forment notamment des espaces de rencontre idéaux pour la prostitution clandestine en raison de leurs emplacements particulièrement passants. Les faits de racolage sont si fréquents qu'il n'est pas rare de les voir relatés. *Le XIXᵉ siècle, journal quotidien, politique et littéraire* propose un article entier à ce sujet en 1892, rapportant comment « les agents chargés de la surveillance faillirent se trouver mal de honte quand ils pénétrèrent dans l'établissement, où ils furent témoins de scènes absolument épiques »[49]. En parallèle des articles, se développe toute une iconographie liant les chalets à la prostitution féminine. L'un de ses meilleurs représentants est Demetrios Galanis, dessinateur presse pour *L'assiette au beurre*, hebdomadaire satirique illustré. Toutefois, il serait faux de postuler que la prostitution ayant cours dans les latrines publiques est exclusivement féminine. Félix Carlier, dans ses *Etudes de pathologie sociale : Les deux prostitutions*, consacre tout un chapitre à la prostitution dite « antiphysique », à savoir masculine. Nous y apprenons que les prostitués appartenant à la typologie des « petits jésus » officient à la manière de « dom juan des latrines publiques »[50]. Ces passes s'accompagnent aussi de larcins, au rang desquels figurent le racket et le vol. En effet, il n'est pas

49. *Le XIXᵉ siècle*, oct. 1892.
50. Carlier, *Etudes de pathologie sociale*, 327.

rare que les petits jésus profitent de la délicate position de leurs clients afin de leur soutirer de l'argent. Dans *Mes lundis en prison*[51], Gustave Macé narre les méfaits d'un petit jésus et de son compère dont la spécialité était de se faire passer pour un inspecteur des mœurs afin d'escroquer des homosexuels dans le désarroi. En parallèle de ces actes de racket, nous dénombrons aussi toute une série de vols ayant pour objet les recettes générées par ces lieux d'aisance. *La petite presse : Journal quotidien*[52] nous apprend en 1887 que quinze chalets de nécessité du quartier Saint-Germain l'Auxerrois ont été dévalisés. Toutefois, il ne s'y produit pas que des délits, mais également des crimes. Les deux infanticides les plus horrifiques de la période se sont déroulés respectivement dans le chalet de nécessité de l'avenue de Breteuil en 1892[53] et dans celui rue de la Contrescarpe en 1897[54]. Enfin, ils constituent un lieu de prédilection pour les suicides. Le journal *La presse* s'est penché sur le rôle des vespasiennes dans les suicides en 1883 : « Frappé comme moi de cette coïncidence, un de mes amis, un fanatique de cette fameuse statistique dont je parlais l'autre jour à cette place, se mit à feuilleter une collection de journaux de 1880 et releva pour Paris seul plus de douze suicides accomplis dans les vespasiennes »[55]. Bien qu'il semble difficile de prévenir ce genre d'agissements, les ingénieurs et la municipalité œuvrent afin de proposer des solutions censées les limiter.

Parmi les innovations proposées, nous observons l'incursion dans les lieux d'aisance d'une loge consacrée à une figure qui ne tardera pas à devenir indissociable de ces édifices : la gardienne. Son apparition s'explique essentiellement par la mixité de ces espaces, pour inciter les dames à y pénétrer et pour veiller à la bonne marche morale des chalets. Le *Cahier des charges de la Concession de cent chalets de nécessité* ne laisse pas planer le moindre doute quant au rôle sécuritaire du gardiennage : « En tout cas, un espace suffisant sera toujours réservé aux communications et au gardiennage, de telle façon que la surveillance et l'entretien ne rencontrent aucun obstacle ». Cette volonté trouve une transcription précoce dans les plans d'ingénieurs. Le tout premier plan d'un chalet de type I dont nous disposons, en date du 10 juin 1874, dressé par Adolphe Alphand propose déjà un large espace pour la loge de la préposée à la surveillance. Par la suite, les ingénieurs des Ponts et Chaussées n'auront de cesse de réinterpréter la taille, mais aussi l'emplacement de la loge. Les ingénieurs tentent de trouver le meilleur emplacement afin d'élaborer un axe optimal pour la surveillance du lieu, tout en garantissant un espace nécessaire pour les différents compartiments d'aisance. Cependant, le maintien de l'ordre au sein des édifices ne saurait être intégralement imparti à

---

51. Macé, *Mes lundis en prison*, 158.
52. *La petite presse*, 30 nov. 1887.
53. *La République du Midi*, 21 oct. 1892.
54. *La presse*, mai 1897.
55. *La presse*, janv. 1883.

des femmes dont la courtoisie et la moralité sont souvent raillées. Henry Gerbault propose un dessin des plus explicites à ce sujet à la une du *Rire : Journal humoristique* le 24 février 1900 (fig. 4), en fusionnant l'image de la gardienne peu affable avec celle de la prostituée de rue. Dès lors, la municipalité fait le choix d'impliquer des gardiens de la paix dans le processus de surveillance. *Le rappel* décrit avec précision ce nouveau système instauré dans les années 1880 :

> Jusqu'à une heure assez avancée de la nuit, la marchande, moyennant une légère gratification, pourra veiller à la propreté du petit édifice. Quand elle fermera boutique, elle sera remplacée par un gardien de la paix qui sera garanti contre les intempéries des saisons par un auvent mobile à volonté. Le gardien sera placé là à poste fixe et relevé de deux en deux heures, durant toute la nuit. Il pourra ainsi surveiller un espace de terrain assez grand et prêter main-forte au premier appel de son voisin ou accourir au premier cri[56].

Cet extrait est d'importance en tant qu'il souligne comment les exigences de moralité engendrent une modification de la forme des édicules par l'ajout, dans le cas présent, d'un auvent mobile, révélant ainsi les intrications entre la municipalité et les ingénieurs dans la gestion des *corps pissants*.

Par ailleurs, la notion de surveillance trouve un autre écho ingénieurial grâce au système d'éclairage mis en place à l'intérieur et à l'extérieur des édicules. L'article 7 du *Cahier des charges de la Concession de cent chalets de nécessité* stipule que « cet éclairage qui devra durer non seulement le même temps que les candélabres de la ville, mais encore tout le reste de la nuit, sera obtenu par plusieurs becs dépensant au minimum par heure 280 litres de gaz pour les chalets de plus grande dimension. Cet éclairage devra se projeter partiellement sur la voie publique par les carreaux supérieurs de chaque édicule ». Cet article nous apprend que ce système repose sur le gaz, dont la maîtrise technique est assurée depuis la fin des années 1830. En plein essor depuis les années 1855, le gaz demeure à la fin des années 1870 un synonyme de la modernité urbaine avant que l'électricité ne commence à faire des émules lors de l'Exposition Universelle de 1881[57]. De plus, ce choix du gaz est une sorte d'évidence en tant qu'il repose sur un réseau de canalisation bien fixé et qui n'a de cesse de se densifier[58]. Ensuite, nous comprenons que l'éclairage, loin d'être chiche, est des plus généreux en reposant sur des becs multiples. Le *Rapport du conseil municipal de Paris* de 1887[59] revient sur le nombre de becs de gaz équipant les chalets de type I, estimant ainsi leur nombre à 693, pour un coût annuel total de 35 360 francs, une coquette dépense qui

---

56. *Le rappel*, févr. 1880.
57. Beltran et Carré, « Une fin de siècle électrique », 92.
58. Williot, « Naissance d'un réseau gazier », 576.
59. *Rapport du conseil municipal de Paris*, janv. 1887.

FIGURE 4 Henry Gerbault (1863–1930). Louis Quinze Centimes, 24 février 1900, Paris. Cote : Bibliothèque nationale de France, département Collections numérisées, 2012–112424.

souligne l'importance de l'éclairage en ces lieux. Enfin, nous saisissons qu'il concourt à celui de l'ensemble de la rue la nuit. Dès lors, les chalets de type I tentent d'assurer l'ordre intérieur, mais aussi extérieur, en se superposant, puis en se substituant aux heures les plus sombres, à la lumière des réverbères à gaz. Nous pouvons néanmoins nous interroger quant à l'efficacité réelle de l'éclairage tant les chalets de type I demeurent des lieux où se déroulent des délits et crimes en tout genre. Il semble alors qu'il importe en réalité peu que cette surveillance soit effective, tant que les signes envoyés aux usagers et riverains simulent son hypothétique efficacité.

Ainsi, l'étude des chalets de type I est informative à trois principaux égards. Tout d'abord, cette typologie s'inscrit autant dans une forme de continuité que de rupture vis-à-vis des urinoirs précédemment implantés. La continuité est lisible par la volonté toujours plus affirmée de dérober le *corps pissant* à la vue des passants. La rupture, quant à elle, s'observe par la forme profondément anti-urbaine des chalets et leur mixité. Pour ces ouvrages à destination des *corps* masculins et féminins, les ingénieurs et les édiles ont tout mis en œuvre afin d'en faire des espaces d'une urbanité recomposée selon les préceptes hygiénistes et moraux de l'époque. Une urbanité qui se heurte toutefois à la mauvaise réputation de ces commodités, engendrant un souci toujours plus marqué de l'ordre et du contrôle appliqués aux *corps pissants* dans la capitale. Si nous avons pu observer comment la ségrégation entre les sexes se met en place dans des édicules mixtes, il nous faut à présent réfléchir à la façon dont la pensée technique a conçu les lieux d'aisance exclusivement réservés aux femmes, à savoir les chalets de type II.

« Là, tout n'est qu'ordre et beauté, luxe, calme et volupté » : Les chalets féminins de type II

Les chalets de type II (fig. 5) apparaissent à Paris en 1879, soit six ans après la création des chalets de type I dans la ville : « De nouveaux chalets de nécessité vont être prochainement édifiés dans Paris sur un modèle analogue à ceux que l'on connaît déjà, place de la Madeleine, square de la Bourse, etc. Ces chalets, d'un modèle uniforme, seront destinés exclusivement aux dames »[60]. Hauts de 3m 30 sur 1m 80 de large[61], ils sont disposés « à quantité égale pour chacun des vingt arrondissements »[62]. Bien que d'apparence similaire aux chalets de type I, tant dans leur forme que leur fonctionnement (fig. 6), ces édicules de type II traduisent une pensée différenciée du *corps* féminin, que nous tâcherons de mettre au jour. Tout d'abord, nous constaterons que cette distinction se forme avant tout autour de la notion de confort. Puis, nous aborderons la thématique de la publicité, dont le rôle est central, tant pour le fonctionnement économique du chalet que dans le processus de formalisation du *corps pissant* féminin. Enfin, nous reviendrons aux inégalités socio-spatiales qu'engendrent ces latrines.

60. *La France*, mai 1879.
61. *La France*, oct. 1879.
62. *La France*, mai 1879.

FIGURE 5 Charles Marville (1813–79). Châlet de nécessité du marché de la cité, v. 1865, Paris. Cote : State Library of Victoria, H2011.126/13.

Au confort des dames : Une pensée différenciée du corps pissant féminin

En découvrant cette typologie, il nous a semblé essentiel de comprendre en quoi elle se distinguait des chalets mixtes. Si la matérialité des édicules demeure inchangée, tout comme le système de gardiennage ou encore la tarification, il convient d'infléchir le regard afin de saisir comment se déploie une pensée ingénieuriale et édilitaire propre aux féminin dans ces paisibles asiles du bonheur vésical.

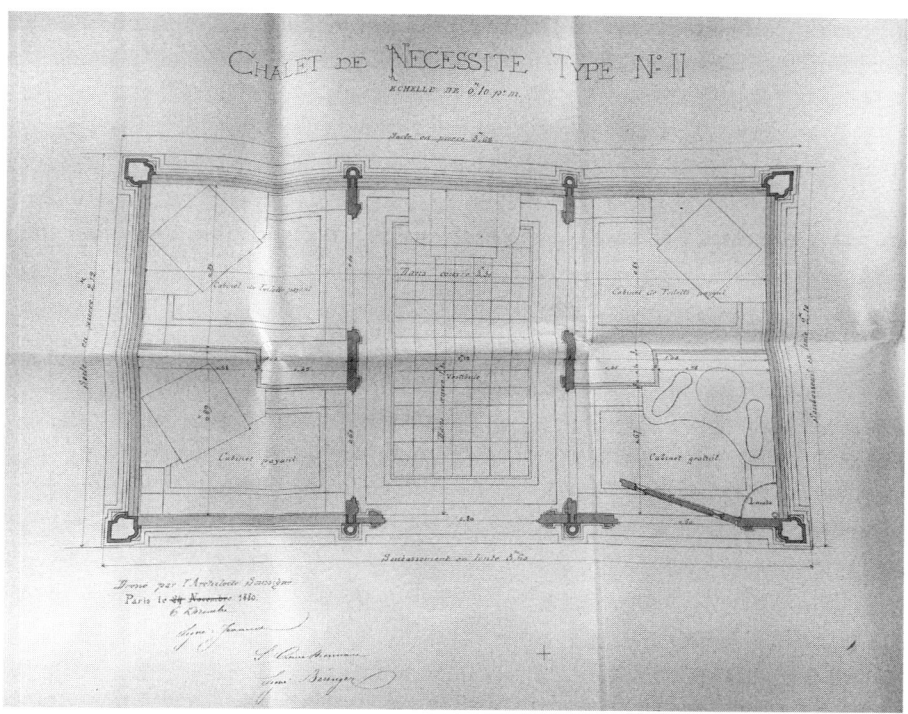

FIGURE 6 Jeannot, architecte. Chalets de nécessité type II, 1880, Paris. Cote : Archives de Paris, VONC 16.

Afin de comprendre cette différence de traitement entre les hommes et les femmes, il convient d'effectuer un bref retour sur la condition féminine au XIXe siècle. L'historienne Yannick Ripa la résume ainsi : « Le XIXe siècle est très dur pour la condition féminine. Empire et Restauration travaillent à inférioriser les femmes, légalisant la hiérarchie des sexes, en la justifiant par des discours scientifiques, philosophiques et religieux »[63]. C'est donc dans ce climat difficile et prompt à la différenciation spatiale[64] entre les genres que germe l'idée d'implanter sur la chaussée des commodités uniquement destinées à la gent féminine. Il y a d'ailleurs urgence, car, rappelons-le, l'interdiction d'uriner dans les rues, maintes et maintes fois rappelée par la préfecture de Police de Paris, s'applique aussi aux femmes, « d'où il résulte que dans cette ville les gendarmes ont fort souvent à rougir et à dresser des procès-verbaux contre des pauvrettes, qu'une cruelle nécessité oblige à outrager la morale publique en s'arrêtant au bord d'un

63. Ripa, *Les femmes, actrices de l'histoire*, 25.
64. Ripa, *Les femmes, actrices de l'histoire*.

trottoir »[65]. Le *corps* féminin apparaît, à cet égard, domestiqué de façon plus précoce que celui des hommes. L'historien Hervé Terral explique ce phénomène par l'éducation qu'elles ont reçue :

> Dans ces éducations où il convient explicitement de se méfier de la séduction et de la tentation, propres à un sexe jugé aussi « beau » que « faible », le corps doit être, d'évidence, « tenu ». Mais cette civilité, si elle requiert des aspects spécifiques, s'inscrit dans une démarche plus générale, celle d'un « processus de civilisation » où Erasme côtoie Jean Baptiste de la Salle et ses Frères des écoles chrétiennes[66].

Plus encore, la dépréciation des parties intimes féminines explicite également le peu d'intérêt porté par la municipalité et par les femmes elles-mêmes à la question des toilettes publiques. Christine Boyer récapitule ainsi la situation : « Dans leur tentative visant à guérir les malaises de la ville, ils [les urbanistes du second Empire] ont nié ses formes physiques, traitant l'espace comme le corps féminin qui, lui aussi au XIXe siècle, était perçu comme le lieu des excès, des hystéries, des maladies et des exclusions »[67]. Néanmoins, cette précarisation du *corps pissant* féminin dans la capitale fait face à des critiques dès la première partie du XIXe siècle. Elles émanent du monde scientifique et médical ainsi que du monde édilitaire. En effet, les médecins de l'époque, bien qu'ils psychopathologisent les utérus, s'intéressent aussi à la bonne santé des vessies et plus généralement des organes reproducteurs, à une période où il est attendu que les femmes enfantent. Nous retrouvons donc de nombreux traités médicaux étudiant la question, comme *Maladies des voies urinaires et des organes génitaux : Préservation et traitement* du Dr Rochon[68] ou encore *Sur la vessie irritable chez la femme (cystopathie hyperhémique)* de Georges Dacheux[69]. Le triomphe de la doctrine hygiéniste, portée par la parole des médecins, chemine donc vers les édiles qui ont tôt fait de se saisir du problème. Pour cause, la municipalité, dans sa lutte épidémiologique contre le choléra, fait face à une amère vérité : les femmes meurent plus que les hommes[70]. Par ailleurs, le monde du travail extra-domestique se féminise à la même époque, en particulier depuis les législations de 1841 et de 1874 restreignant le travail des enfants[71]. Elles sont donc plus visibles sur la chaussée et ces mobilités pendulaires rendent nécessaires des moments et des lieux de soulagement. Mobilités pendulaires qui sont d'ailleurs prises en compte dans les

---

65. *Le moderniste illustré*, juin 1889.
66. Terral, « Le corps de la jeune fille », 103.
67. Boyer, *City of Collective Memory*, 18.
68. Rochon, *Maladies des voies urinaires*.
69. Dacheux, *Sur la vessie irritable chez la femme*.
70. *Journal de la société statistique de Paris*, « Note statistique sur le choléra de 1832, 1849 et 1854 ».
71. Ripa, *Les femmes, actrices de l'histoire*.

logiques d'implantation des chalets comme pour le cas de la place de la Bourse et de son célèbre bureau de la Compagnie des Omnibus, par où transite le tout Paris des travailleurs et travailleuses. Si tout est fait pour que les usagères ne soient pas dérangées, subsiste toujours le fait qu'elles sont visibles lorsqu'elles rentrent dans un water-closet. Action anodine pour un observateur du XXIᵉ siècle, elle devient un véritable spectacle pour celui du XIXᵉ siècle. A la lecture du *Journal amusant : Journal illustré, journal d'images, journal comique, critique, satirique, etc.*, nous pourrions presque parler de voyeurisme : « Ce que je voudrais voir, c'est la première audacieuse qui pénétrera dans ce paisible asile. Il lui faudra certainement une intrépidité exceptionnelle car il y aura bien un millier de spectateurs pour la regarder ! »[72]. C'est pourquoi les chalets de type II, exclusivement conçus pour les femmes, tentent de dépasser cette hostilité préliminaire en faisant de ces lieux les chantres du confort vésical de ces dames.

Avant d'étudier les différentes modalités du confort dans les chalets pour dames, il convient de revenir sur la signification de ce terme. Si dans son sens ancien, employé jusqu'au XVIIᵉ siècle, *confort* signifie aide, assistance et secours, ce n'est qu'à partir de 1842, par l'entremise de la définition de l'Académie, qu'il devient le synonyme du bien-être matériel. Cette nouvelle définition est le fruit de la transposition du sens britannique du concept de confort tel qu'il s'est développé à partir du XVIIIᵉ siècle. Par *confort*, on entend également entre les années 1840 et 1850 « la maniabilité des objets, le plus grand nombre de services qu'ils rendent »[73]. En parallèle germe une vision du confort, portée par des auteurs tels que Victor Considérant, supposant la totale redistribution des fluides, mais aussi leur instrumentalisation et leur mécanisation[74]. Nous pouvons à présent nous pencher sur la transcription de la notion de confort au sein de ce type d'édicule. Tout d'abord, les chalets de type II s'inscrivent dans une forme de continuité avec la définition ancienne du confort, en tant qu'ils fournissent un asile aux usagères, s'apparentant alors à un « confort-réconfort »[75]. Le sens anglosaxon est, lui aussi, relevé par la presse qui voit dans ces édicules « une coquetterie toute française et un confort tout anglo-saxon »[76]. Notons au passage que si cette notion ne transparaît pas dans les termes de *chalet* ou encore de *lavatory*, elle marquera l'appellation des premières toilettes publiques installées aux Etats-Unis, nommées « comfort stations ». Toutefois, c'est principalement autour de l'hybridation des fonctions que le confort se matérialise. Par opposition au chalet de type I qui ne propose que des cabinets d'aisance, ceux-là offrent à leurs clientes

---

72. *Le journal amusant*, nov. 1882.
73. Vigarello, « Confort et hygiène en France au XIXᵉ siècle », 93.
74. Vigarello, « Confort et hygiène en France au XIXᵉ siècle ».
75. Goubert, *Du luxe au confort*.
76. *Le XIXᵉ siècle, journal quotidien, politique et littéraire*, sept. 1891.

des cabinets de toilette. La différence repose sur la présence d'un lavabo dans les cabinets de toilette, permettant aux femmes d'effectuer un brin de toilette supplémentaire[77]. Cet ajout fait sens à une époque où la propreté devient un gage de respectabilité, en particulier pour les femmes : « Etre propre, c'est avoir de la tenue, aux deux sens du terme ».[78] Cependant, la multiplicité des services proposée aux femmes ne s'arrête pas là. L'édition du 28 juillet 1872 du *Figaro* nous apprend que « chaque chalet comprendrait trois compartiments, dans ces compartiments seraient placés un miroir, une boîte contenant des épingles, et quelques bons livres »[79]. Nous supposons que le miroir et les épingles permettent aux usagères d'ajuster leur coiffure avant de sortir des édicules. Il est important de souligner que les chalets de type II miment le confort intérieur des demeures bourgeoises inférant lui-même une conception de la place et du rôle des femmes. Il est d'ailleurs essentiel de noter que l'adjectif « confortable » fait l'objet d'une mention dans le *Cahier des charges de la Concession de cent chalets de nécessité* au sujet de l'intérieur des chalets pour dames : « L'intérieur des chalets sera confortable ; les cabinets d'aisance auront des sièges en chêne ciré avec une cuvette en porcelaine à bouchon et effet hydraulique. La garniture des cabinets de toilette sera en marbre et porcelaine ». Ces éléments indiquent que le confort, tel que perçu par les édiles et les ingénieurs, est celui de la redistribution et de la mécanisation progressive des fluides, comme l'indique le siphonnage à pression hydraulique, procédé par lequel les femmes peuvent pratiquer des ablutions corporelles partielles au sein des édicules. De plus, la présence du marbre, tout comme dans les chalets mixtes et surtout celle de la porcelaine, trace une filiation entre la mécanisation et l'hygiénisme. En effet, loin d'être anodins, ces deux matériaux, par leurs propriétés nettoyantes et par leur couleur blanche, en particulier à l'intérieur des chalets sont tributaires de la vision hygiéniste de l'époque. Le rôle déterminant des couleurs dans les processus de guérison et de salubrité est longuement étudié par Florence Nightingale dans *Notes on Nursing : What Nursing Is, What Nursing Is Not* publié en 1860 et traduit en français dès 1862, attestant la popularité de cet ouvrage. Elle écrit à ce propos : « Nous savons peu de choses sur la manière dont nous sommes affectés par la forme, la couleur et la lumière, mais nous savons qu'ils ont un effet réel et physique. La variété des formes et la vivacité des couleurs sur les objets présentés aux patients sont un moyen de récupération réel »[80]. La pensée hygiéniste du soin, centrée autour du rôle salvateur de la couleur, opère donc un transfert du milieu hospitalier vers

---

77. *La petite presse*, avr. 1881.
78. Perrot, *Le travail des apparences*, 107.
79. *Le Figaro*, juil. 1872.
80. Nightingale, *Des soins à donner aux malades*, 109.

celui du mobilier urbain d'aisance. En outre, le terme *porcelaine* finira même par désigner par métonymie les toilettes publiques[81]. Ainsi, les différentes acceptations du soin dépassent le simple cadre discursif en participant à l'aménagement des chalets pour femmes, en plaçant cette notion au cœur de la réforme du *corps* féminin. Une réforme qui se voit compléter par la publicité.

Au bonheur des dames : Retour sur la question de la publicité

La publicité, si elle a su trouver dans le mobilier urbain de nouveaux murs pour l'affichage, revêt également un double rôle normatif dans les chalets. Nous verrons d'abord comment elle fait l'objet d'un encadrement et d'une surveillance, puis nous montrerons en quoi elle participe à la fabrique d'une certaine idée de la féminité.

Souligné par Roxane Bonnardel-Mira, « le jeu entre affichage licite et affichage illicite apparaît dans les usages mêmes du mobilier urbain »[82] et tout particulièrement dans les commodités réservées aux dames. Longtemps demeuré anarchique, l'encadrement de l'affichage tend à se formaliser à partir de la loi du 29 juillet 1881. Ainsi, dès 1882, les concessionnaires sont tenus de soumettre à l'ingénieur, chef du service de l'Eclairage, des Promenades et des Plantations le contenu des réclames qu'ils souhaitent apposer sur les façades extérieures des édicules, comme en attestent les soixante et un billets conservés aux Archives de Paris sous la cote Vo3 425. De plus, la réglementation concerne pareillement les matériaux des affiches : dès 1889, les affiches en papier sont supprimées au profit de celles en tôle émaillée. Là encore, cette interdiction est à lire sous le jour de la destination féminine de ces lieux. En effet, les affiches en papier facilitent les inscriptions obscènes dont les usagères se plaignent. A ce titre, Ernest Camescasse, préfet de police de 1881 à 1885, adresse le 31 juillet 1883 un courrier à l'ingénieur chef M. Bartet afin de le sommer de régler dans les plus brefs délais les problèmes liés aux inscriptions et dessins pornographiques des chalets du concessionnaire Béranger[83]. Ne parvenant hélas pas à endiguer ce phénomène, la municipalité est contrainte de supprimer l'affichage à l'intérieur des cabines à partir de 1889. Nous saisissons ainsi comment les exigences de pudeur et de moralité que la municipalité calque sur le *corps pissant* féminin aboutissent à une normalisation croissante des pratiques d'affichage dans les toilettes pour femmes.

Avant de revenir sur la manière dont les différents types de publicités proposés aux dames influent sur les conceptions et la perception de la féminité dans le dernier tiers du XIXe siècle, il nous semble important de mettre l'accent sur la

---

81. Routh, *Guide Porcelain*.
82. Bonnardel-Mira, « Nouvelles pratiques de l'affichage commercial », 190.
83. Archives de Paris, VONC 16.

matérialité de ces derniers. Lorsque nous songeons aux affiches produites à cette période, nous imaginons de grandes chromolithographies, popularisées par Jules Chéret[84] et magnifiées par des artistes tels que Toulouse-Lautrec ou encore Mucha. Si les photographies d'époque ne laissent aucun doute quant à la présence d'affiches illustrées à l'extérieur des chalets, il ne faut pas pour autant sous-estimer la présence d'affiches comportant simplement du texte. Comme le rappelle Marc Martin : « Une partie seulement des affiches sont illustrées ; beaucoup ne comportent toujours qu'un texte, car l'affiche illustrée coûte cher »[85]. Par-delà la question du coût, ces affiches énigmatiques, ne comportant parfois guère plus qu'un mot, sont très efficaces : « Il suffit qu'un mot, énorme, coure le long des murs. Le promeneur, obsédé, trouve en rentrant chez lui, dans son journal, le commentaire insinuant du mot hurleur »[86]. Indissociable du renouvellement du paysage parisien comme cadre figuratif, nous retrouvons de nombreuses descriptions dithyrambiques à ce sujet de la part d'écrivains, à l'instar de Joris-Karl Huysmans[87].

Néanmoins, le paysage ne se contente pas d'être un cadre figuratif, il constitue aussi une norme moralisatrice. C'est au prisme de cette double perspective qu'il convient d'étudier ces publicités. Comme l'ont rappelé Anne-Sophie Aguilar et Eléonore Challine : « La publicité, en plus de s'imposer dans l'environnement matériel des individus devenus des consommateurs, cherche donc à en pénétrer les esprits »[88]. Dès lors, il est impératif de se pencher sur les types de produits proposés aux dames. A l'aide des soixante et un billets mentionnés en amont, il nous a été possible d'isoler six principales catégories : divertissement, esthétique, mode, domotique, alimentation et produits pharmaceutiques. La partie divertissement se focalise essentiellement autour des bals (Bal du Moulin de la Galette) et des théâtres (Théâtre lyrique populaire—Etienne Marcel). Les réclames concernant l'esthétique font la part belle aux coiffeurs (Bétry Delval—Coiffeur) et à la parfumerie (Eau de toilette au maté). La mode est indissociable des grands magasins (Grands Magasins de la Samaritaine, Grand bazar de l'Hôtel de Ville, etc.). La partie domotique s'axe autour de la couture (Machine à coudre Singer, Lampe spéciale pour machine à coudre) et du « confort » domestique (Allume-feux écossais, Nouvelle cuisinière universelle). Au sujet de l'alimentation, ce sont les cafés et boissons à la mode qui sont promus (Café Fouquet, Chocolat du planteur). Enfin parmi les produits pharmaceutiques, nous notons une préoccupation croissante pour l'hygiène bucco-dentaire (Crème dentifrice

---

84. Martin, *Trois siècles de publicité en France*.
85. Martin, *Trois siècles de publicité en France*, 90.
86. Uzanne, *Félix Vallotton*, 70.
87. Huysmans, *Certains*.
88. Aguilar et Challine, « Introduction », 15.

antiseptique, Dentifrice au sumac), pour l'hygiène corporelle (Savonnerie du Cosmydor, Bains de Bréda) ainsi que des remèdes aux maladies « féminines » (Maladies secrètes devient Cabinet du Dr. Emmanet maladie du sang et de la peau, Docteur Armand—traitement spécial des vices du sang et maladies des femmes). Ces catégories esquissent un modèle socioculturel bourgeois révélateur de la place des femmes dans la société parisienne du XIXe siècle. Ainsi, les réclames concernant l'esthétique et la mode réaffirment la nécessaire coquetterie attendue des femmes aisées de la période, un idéal auquel tendent également les moins fortunées qui utilisent ces lieux d'aisance par désir mimétique. Celles au sujet de la domotique et de l'alimentation font écho aux rôles de maîtresse de maison et de ménagère qui incombent aux bourgeoises et aux femmes des classes populaires. Les affiches relatives au divertissement sont le témoin d'une formation d'une culture du loisir urbain au XIXe siècle dans laquelle les femmes ont un rôle à jouer et une place à tenir. Comme le rappellent Robert Beck et Anna Madœuf : « A l'époque où le salon et la maison constituaient les lieux essentiels de la socialité, le théâtre était, à l'extérieur, un espace où les femmes jouaient un rôle de premier plan »[89]. Nous comprenons alors comment, les contenus et les « représentations que les publicitaires donnent de la femme procèdent, en effet, par standardisation et exagération d'expressions et de comportements qui, dans la réalité, sont déjà fortement ritualisés »[90]. Ainsi, la publicité présente dans les chalets de type II illustre à quel point le *corps pissant* féminin constitue un support idéal afin de parfaire la formation par la consommation à une féminité normée et encadrée, tout en esquissant un territoire de possible mobilité pour les femmes par l'entremise des voyages et du loisir. Plus encore, « les pratiques de l'affichage influencent la culture urbaine entendue comme les pratiques culturelles, rapports sociaux et représentations façonnés dans l'espace urbain, et qui participent en retour à son évolution »[91], expliquant de ce fait la prohibition des affiches à caractère médical vers la fin du siècle, en raison de leur inadéquation avec la pensée morale et salubre de la ville. Si elles connaissent un franc succès durant l'ensemble du XIXe siècle, elles tombent en disgrâce au début du XXe siècle. Ce désaveu s'explique pour trois raisons principales. Tout d'abord, les intitulés des publicités sont jugés mensongers, en particulier lorsqu'ils mentionnent l'adjectif « secret ». Si le remède dit secret est « une formidable innovation commerciale destinée à briser net les hésitations du consommateur »[92], cette appellation est jugée « fallacieuse

---

89. Beck et Madœuf, *Divertissements et loisirs*, 121.
90. Goffman, « La ritualisation de la féminité », 48.
91. Bonnardel-Mira, « Nouvelles pratiques de l'affichage commercial », 182.
92. Sueur, « Les pharmaciens et la médiatisation », 38.

et indigne du corps médical »[93] par la Société française de prophylaxie sanitaire et morale. Le deuxième argument est celui du respect de la science médicale. Le bulletin de la Société française de prophylaxie sanitaire et morale est plus que disert à ce propos : « Au point de vue de la science, ces affiches constituent un scandale et un abus : un scandale, parce qu'il est intolérable de voir jeter le discrédit sur le seul médicament spécifique que nous ayons à opposer à la plus redoutable des maladies [. . .], mais surtout [cette publicité] constitue un danger permanent pour la santé publique »[94]. Il est intéressant de noter le glissement dans l'échelle de dangerosité des maladies. Si au début du XIXe siècle, la maladie la plus crainte était le choléra, relevons qu'en ce début de XXe siècle, c'est de la syphilis dont les édiles et médecins ont le plus peur, signalant ainsi la bascule entre l'hygiène publique et l'hygiène sociale. C'est à l'aune de ce changement que la troisième raison invoquée prend tout son sens : il s'agit de la morale. A ce titre, ce qui choque le plus les auteurs du bulletin est la présence des affiches sur les parois des cabines de chalets de nécessité pour dames, les exposant à des réclames proposant « la guérison des maladies des femmes en général, et en particulier des écoulements invétérés »[95]. En définitive, c'est parce qu'elles sont présentes dans les toilettes pour femmes que ces réclames sont supprimées. Dès lors, il résulte de la pensée appliquée au *corps* des féminin dans la ville, une modification du paysage publicitaire, mais aussi morale dans Paris. Ainsi, l'observation des pratiques publicitaires au sein des chalets de type II participe à la meilleure compréhension matérielle des édicules, tout en soulignant comment la régulation des affiches suggère la place singulière attribuée au *corps pissant* féminin qui se définit par et pour la publicité. Néanmoins, si nous avons observé la dimension bourgeoise des water-closets, il nous faut à présent nous pencher sur la variété des profils socio-économiques des usagères de ces lieux.

Une « discrétion » accessible à toutes ?

Elégants, confortables et sécurisés, les chalets de type II ont su gagner une place de choix au sein de la capitale en moins de trois décennies. Par ailleurs, nous avons aussi insisté sur le caractère bourgeois de la normativité de ces lieux, fait d'ailleurs souvent repris par le dessin de presse, comme en atteste celui de Haye dans *Le journal pour tous* du 6 octobre 1904, montrant une gardienne confrontée à une cliente vêtue comme une bourgeoise dont elle raille l'odeur : « Brrrrr ! Cette petite empeste l'eau de Cologne ! » (fig. 7). Pourtant, toutes les femmes, indifféremment de leur classe sociale doivent faire leur besoin. Il s'agira donc d'étudier

93. *Bulletin mensuel*, mai 1902.
94. *Bulletin mensuel*, mai 1902.
95. *Bulletin mensuel*, mai 1902.

la façon dont se manifeste ingé-
nieurialement, économique-
ment et spatialement des distinctions entre les *corps pissants*
féminins dans le Paris de la
fin de XIXᵉ siècle. Nous reviendrons d'abord sur la pensée
sociale des ingénieurs, obser-
vable via les cabines gratuites.
Puis, nous nous pencherons sur
les tarifs pratiqués dans ces
espaces. Enfin, nous étudierons
la répartition socio-spatiale des
chalets de type II dans Paris.

Grands œuvres des ingé-
nieurs, les chalets de nécessité
pour dames sont les témoins de
la mise en œuvre d'une forme
de pensée sociale dans la capitale. En effet, cette dimension
chère au cœur d'Antoine Picon
éclaire les différents travaux réa-
lisés par ce corps de métier au
XIXᵉ siècle : « Ils se prenaient
à rêver d'une société différente qui saurait conjuguer plus
efficacement que celle qu'ils
avaient sous les yeux le progrès

AU CHALET DE NÉCESSITÉ

LA GARDIENNE. — Brrrrr!!... cette petite empeste
l'eau de Cologne.
Dessin de **Haye.**

FIGURE 7 Lucien Haye (1873–1946). Au Chalet de
nécessité, octobre 1904, Paris. Cote : Bibliothèque
nationale de France, département Littérature et art,
FOLZ-652.

matériel et l'amélioration des conditions de vie des différentes classes sociales, le
prolétariat urbain en particulier »[96]. Dans les chalets pour dames, le progrès
matériel, ayant pour visée l'amélioration des conditions de vie, trouve un écho
dans la cabine gratuite dont chaque édicule doit s'équiper. Elle se différencie
néanmoins des autres. Tout d'abord, il s'agit d'un simple cabinet d'aisance
sans lavabo. Ensuite, il est de taille plus modeste que ceux des autres catégories.
Enfin, nous y pénétrons par une entrée spéciale, entraînant une différenciation
spatiale au sein de ces édicules : « Les cabinets gratuits pourront avoir des
entrées spéciales à l'extérieur, sous réserve de l'établissement de sonnerie

96. Picon, « La pensée sociale », 28.

d'appel correspondant avec le bureau de la gardienne »[97]. Bien que conçues pour rendre service aux plus modestes, ces cabines induisent une différence de traitement entre les *corps pissants* féminins selon leur appartenance sociale. Nous sommes d'ailleurs en droit de nous demander si ces cabinets sont réellement utilisés, tant les ingénieurs sont contraints à livrer bataille aux concessionnaires. Agacé par le refus du concessionnaire M. Dorion, l'ingénieur chef C. Montez adresse une note en 1879[98] pour remédier à la situation. Dans cette note, il réaffirme la nécessité de « penser à la population peu aisée », tout en fustigeant « les idées d'exclusion du public non-payant » de M. Dorion. Se mettent alors en place des stratégies afin d'éviter que ces cabines ne soient utilisées. Dans la note, on apprend que le concessionnaire « a pris soin de les tenir constamment fermés », par exemple. Bien que ces manœuvres fassent l'objet de réprimande, nous remarquons que cette règle est le plus souvent ignorée. Si les concessionnaires apparaissent réfractaires, c'est en raison du manque à gagner. Il est d'importance de rappeler que ces derniers tirent leurs revenus de l'affichage publicitaire et des coûts d'entrée qui évoluent durant l'ensemble de la période. En 1879, le coût est de « 15 c. pour les cabinets d'aisance, et à 25 c. pour ceux de toilette »[99]. En 1881, les prix d'accès ont quelque peu été revus à la baisse, variant « de 15 à 5 centimes, selon que l'on demande un cabinet complet (avec lavabo) ou simple, de l'une ou l'autre espèce »[100]. Afin de saisir au mieux ce à quoi correspond une telle dépense, nous avons recherché le prix du kilogramme de pain à la même époque. Selon le *Journal de la société statistique de Paris*[101], il s'élève à 42,7 centimes pour l'année 1881. Ainsi, utiliser un chalet de nécessité de type II, revient à dépenser l'équivalent d'un tiers ou d'un neuvième du prix d'un kilogramme de pain, une dépense impossible pour les populations les plus précaires. A ces inégalités socio-économiques se superposent des inégalités socio-spatiales. Pourtant, une forme d'égalité spatiale semblait initialement habiter le projet d'installation des chalets de type II, comme le stipule le quotidien *La France : Politique, scientifique et littéraire* : « Ces chalets pourraient contenir, comme ceux que nous citons plus haut, un cabinet de toilette à la disposition des dames qui voudraient payer un supplément. Sous ces réserves et sous celle du choix des emplacements que la Ville se charge de fixer à quantité égale pour chacun des vingt arrondissements »[102]. Dans les faits, il n'en est hélas rien. Sur la quarantaine de modèles existants, la liste des emplacements, fournie par la Société anonyme des Chalets de

97. Archives de Paris, VONC 17.
98. Archives de Paris, VONC 1262.
99. *La France*, oct. 1879.
100. *La petite presse*, avr. 1881.
101. Proust, « Le prix du kilogramme de pain ».
102. *La France*, mai 1879.

nécessité en 1885,[103] nous apprend qu'il n'y en a qu'un seul dans le XVIIIᵉ, rue de Clignancourt, un seul dans le XVIIᵉ, au niveau de l'avenue de Clichy, et deux dans le IIIᵉ arrondissement vers le square des Arts et Métiers. Nous n'en retrouvons pas non plus dans les XVᵉ, XIIIᵉ, XIXᵉ et XXᵉ arrondissements qui, tout comme les XVIIᵉ, XVIIIᵉ et IIIᵉ arrondissements concentrent les populations les plus pauvres de Paris. Ces commodités sont donc le lieu privilégié d'un entre soi social, ne favorisant donc que peu la mixité sociale, allant ainsi à l'encontre de sa vocation première, à savoir fournir un espace décent et salubre à toutes.

Ainsi, l'étude des chalets de type II est capitale afin d'appréhender comment ces lieux incorporent tant du point de vue de leur conception que de leur décoration intérieure les différentes pensées du soin à l'œuvre dans la dernière partie du XIXᵉ dans le but de contribuer à la définition singulière du *corps pissant* féminin dans la capitale, centrée autour de la notion séminale de confort, mais aussi de discrétion, de pudeur et de salubrité. Plus encore, ce mobilier de rue accompagne la formation d'une certaine idée de la féminité à laquelle les corps féminins se confrontent par l'entremise des affiches publicitaires. En outre, il interroge l'accessibilité de ces lieux tant d'un point de vue technique qu'économique et spatial à l'ensemble des Parisiennes. Il apparaît alors nettement que le *corps pissant* féminin se conjugue au pluriel, faisant ainsi entrevoir un certain nombre d'inégalités de traitement pour des raisons socio-économiques entre les usagères.

Cet article, consacré aux chalets de nécessité de type I et II, permet de saisir en quoi cette typologie nouvelle et monumentale du mobilier de rue, apparue dans les années 1870, marque un cran supplémentaire dans la volonté édilitaire et ingénieuriale de dérober les *corps pissants* masculins et féminins à la vue des passants, se plaçant de ce fait dans la continuité des travaux entrepris sur les urinoirs à la même époque. Nous avons d'abord essayé de montrer comment l'architecture des lieux participe de cette dissimulation ainsi qu'à sa légitimation. L'engouement pour l'architecture alpine contribue à la fois à offrir un abri sûr pour l'ensemble des usagers, tout en accentuant la dimension propre et salubre des lieux. Bien que ces édicules réactivent une forme d'urbanophobie, ils se parent d'un certain nombre d'atours réaffirmant la dimension urbaine et surtout moderne de ces commodités publiques. Nous pouvons à ce titre citer l'emploi de matériaux comme le fer ou le verre, le recours au gaz ou encore le soin apporté à l'évacuation presque systématique des matières au tout-à-l'égout. Toutefois, il n'aurait pas été possible d'appréhender pleinement la préoccupation croissante de discrétion associée à l'assouvissement de ces besoins dans l'espace public, sans évoquer la destination féminine de ce mobilier de rue. En effet, c'est en partie parce qu'ils s'adressent aux dames que les chalets de type I et de type II arborent cette forme particulière. De plus, leur présence nécessite à

103. Archives Paris, Vo3 427.

la fois une ségrégation interne au lieu ainsi qu'un système efficace de surveillance, incarnée dans la figure tutélaire de la gardienne. Non contents de renouveler le paysage figuratif de la capitale, les chalets et surtout ceux de type II imposent au pavé et aux corps s'y trouvant des exigences moralisatrices recomposées. A dominante bourgeoise, ces préceptes incluent la discrétion, mais aussi une certaine idée du confort. Cette dernière explique en partie l'agencement intérieur des lieux, fondé sur le luxe et la praticité des matériaux ainsi que sur la poly-fonctionnalité des cabines, permettant d'uriner, de se nettoyer ou encore de se coiffer. Le confort comme norme et idéal de vie est d'ailleurs véhiculé par la publicité extérieure et intérieure de ces commodités, rappelant sans cesse les dames à leur rôle de maîtresse de maison, de coquette et de frivole. Néanmoins, si les chalets élaborent une pensée différenciée entre les *corps pissants* masculins et féminins, ils reconduisent aussi un certain nombre d'inégalités socio-économiques et socio-spatiales. Onéreux et trop peu présents dans les quartiers populeux et industrieux, ils ne sont pas accessibles aux femmes aux bourses les plus modestes. Elles devront attendre les années 1890 afin de bénéficier des urinoirs gratuits pour dames, modèle Doriot. A l'heure de la conclusion surgit une ultime question : quel devenir pour les chalets ? Bien qu'ils perdurent globalement jusque dans les années 1930, avant d'être détruits ou réaffectés, leur popularité commence déjà à se tarir au début du XXe siècle[104]. En effet, nombreux sont confrontés à l'irrésistible ascension du métropolitain. Trop encombrants et souvent situés dans les mêmes axes constructifs que les futures lignes de métro, les propositions de déplacement des édicules sont légion. Loin d'être marginale, cette série de déplacement et de suppression annonce l'essor de la pensée du souterrain et surtout de la vie souterraine. Une vie dans laquelle s'inséreront à partir des années 1900 les toilettes pour dames sous la forme de *lavatories* souterrains, symptomatique d'un énième aller-retour entre les pratiques ingénieuriales françaises et britanniques.

LOUISE THIROUX prépare une thèse intitulée « Du raccommodage au soin : Penser Paris par son mobilier urbain " de confort " (XIXe-XXe siècles) », sous la direction d'Emilie d'Orgeix à l'Ecole Pratique des Hautes Etudes–Université PSL. Elle est également chargée d'études et de recherche à l'Institut national d'histoire de l'art pour le programme *Richelieu, Histoire du quartier*.

LOUISE THIROUX is preparing a thesis titled « Du raccommodage au soin : penser Paris par son mobilier urbain " de confort " (XIXe–XXe siècles) », under the supervision of Emilie d'Orgeix at the Ecole Pratique des Hautes Etudes–Université PSL. She is also in charge of studies and research at the Institut national d'histoire de l'art for the *Richelieu, Histoire du quartier* program.

104. Simon, « Une histoire des lieux d'aisances publics à Paris ».

Remerciements

L'auteur remercie les relecteurs anonymes de la revue *French Historical Studies* ainsi qu' April Shelford et Peter Soppelsa, les deux éditeurs de ce numéro spécial, mais aussi l'Ecole Pratique des Hautes Etudes–Université PSL, Emilie d'Orgeix et l'Institut national d'histoire de l'art pour leurs encouragements.

The author thanks the anonymous reviewers of *French Historical Studies* and April Shelford and Peter Soppelsa, the two guest editors of this special issue. The author would also like to thank Ecole Pratique des Hautes Etudes–Université PSL, Emilie d'Orgeix and Institut national d'histoire de l'art for their encouragement.

Références

Aguilar, Anne-Sophie, et Eléonore Challine. « Introduction : Tout le monde déteste la publicité ? » *Sociétés et représentations*, n°54 (2022) : 10–18.

Assainissement de Paris : Résolutions votées par la commission technique de l'assainissement de Paris et résumé des travaux de la commission. Chaix, 1883.

Baudrillard, Jean. *Le système des objets*. Gallimard, 1968.

Beck, Robert, et Anna Madœuf. *Divertissements et loisirs dans les sociétés urbaines à l'époque moderne et contemporaine*. Presses universitaires François-Rabelais, 2005.

Becquerel, Alfred. *Traité élémentaire d'hygiène privée et publique*. Emile Beaugrand, 1877.

Beltran, Alain, et Patrice A. Carré. « Une fin de siècle électrique ». *Les cahiers de médiologie*, vol. 10, n° 2 (2000) : 90–101.

Benjamin, Walter. *Paris, capitale du XIXe siècle*. Allia, 2003.

Bonnardel-Mira, Roxane. « Nouvelles pratiques de l'affichage commercial et recomposition de la culture urbaine parisienne à la fin du XIXe siècle ». *Revue d'histoire du XIXe siècle*, n°63 (2021) : 178–85.

Boyer, Christine. *The City of Collective Memory : Its Historical Imagery and Architectural Entertainments*. MIT Press, 1996.

Carlier, Félix. *Etudes de pathologie sociale : Les deux prostitutions*. E. Dentu, 1887.

Dacheux, Georges. *Sur la Vessie irritable chez la femme (cystopathie hyperhémique)*. Steiheil, 1894.

Daly, César. *Revue générale de l'architecture et des travaux publics*. Ducher, 1878.

Delamare, Nicolas. *Traité de la police*, t. 1. Jean-Pierre Cot, 1705.

Giedion, Sigfried. *Construire en France, construire en fer, construire en béton*. Editions de la Villette, 2000.

Goffman, Erving. « La ritualisation de la féminité ». *Actes de la recherche en sciences sociales* 14 (1977) : 42–56.

Goubert, Jean-Pierre. *Du luxe au confort*. Belin, 1988.

Haussmann, Georges-Eugène. *Mémoires*, t. 3. Havard, 1890.

Huysmans, Joris-Karl. *Certains : G. Moreau, Degas, Chéret, Whistler, Rops, le Monstre, le Fer, etc.* Tresse et Stock, 1889.

Laroulandie, Fabrice. « Les égouts de Paris au XIXe siècle : L'enfer vaincu et l'utopie dépassée ». *Cahiers de Fontenay*, n°69–70 (1993) : 120–32.

Lemoine, Bertrand. *L'architecture du fer : France, XIXe siècle*. Champ Vallon, 1986.

Lévy, Albert. « L'intégration de la santé dans l'éco-urbanisme : " Prendre soin du territoire " ». *Scienze del territorio* 10, n°1 (2022) : 108–15.

Loyer, François. *Paris XIXe siècle : L'immeuble et la rue*. Hazan, 1994.

Macé, Gustave. *Mes lundis en prison*. [s.n.], 1889.

Maillard, Claude. *Les vespasiennes de Paris ; ou, les Précieux édicules*. La jeune parque, 1967.

Martin, Marc. *Trois siècles de publicité en France*. Odile Jacob, 1992.

Nightingale, Florence. *Des soins à donner aux malades : Ce qu'il faut faire, ce qu'il faut éviter*. Didier et Cie, 1862.

Perrot, Philippe. *Le travail des apparences ; ou, Les transformations du corps féminin, XVIIIe–XIXe siècles*. Seuil, 1984.

Picon, Antoine. « La pensée sociale et politique des ingénieurs des Ponts et Chaussées ». *Pour mémoire*, n°18 (2016) : 18–37.

Picon, Antoine. *L'ornement architectural : Entre subjectivité et politique*. Presses polytechniques et universitaires romandes, 2016.

Proust, Pierre. « Le prix du kilogramme de pain à Paris de 1880 à 1936 ». *Journal de la société statistique de Paris*, t. 78 (1937): 100–108.

Ripa, Yannick. *Les femmes, actrices de l'histoire : France, de 1789 à nos jours*. Armand Colin, 2010.

Rochon, M. *Maladies des voies urinaires et des organes génitaux : Préservation et traitement*. André Sagnier, 1866.

Rosenberg, Harold. *La tradition du nouveau*. Editions de Minuit, 1962.

Routh, Jonathan. *The Guide Porcelain : The Loos of Paris*. Wolf Publishing, 1966.

Simon, Miriam. « Une histoire des lieux d'aisances publics à Paris ». In Situ [En ligne], n°51 (2023).

Sueur, Nicolas. « Les pharmaciens et la médiatisation de la spécialité au XIXe siècle ». *Le temps des médias* 23, n°2 (2014) : 22–40.

Terral, Hervé. « Le corps de la jeune fille : Un invariant pédagogique au fil des siècles ? ». *Le Télémaque*, vol. 25, n°1 (2004) : 92–108.

Uzanne, Octave. *Félix Vallotton, Les Rassemblements : Physiologies de la rue*. H. Floury pour les Bibliophiles indépendants, 1896.

Vernes, Michel. « Le chalet infidèle ; ou, Les dérives d'une architecture vertueuse et de son paysage de rêve ». *Revue d'histoire du XIXe siècle*, n°32 (2006) : 110–24.

Vigarello, Georges. « Confort et hygiène en France au XIXe siècle ». Dans Goubert, *Du luxe au confort*, 88–96.

Williot, Jean-Pierre. « Naissance d'un réseau gazier à Paris au XIXe siècle : Distribution gazière et éclairage ». *Histoire, économie et société*, n°4 (1989) : 558–76.

Automobiles, Entrepreneurs, and Empire
Auto-Tourism in Interwar French North Africa

ZOHAR SAPIR DVIR

ABSTRACT   By the end of World War I, business leaders began promoting the automobile as the ideal technology for touring French North Africa. Leading industrial companies such as Dunlop, Michelin, the Compagnie Générale Transatlantique, and Citroën led initiatives to promote leisured automobile travel in Morocco, Algeria, and Tunisia. These efforts aligned with the French colonial administration's increasing focus on technology, especially motorized mobility to develop, maintain, and govern its North African empire. This article explores the intersection of technology and colonialism in French North Africa during the interwar period through the lens of automobile-based tourism. Questioning the divisions between politics, technology, and leisure, it examines the ideological uses and material effects of auto-tourism from the perspective of mobility. In doing so, it highlights the role of leisure-driven colonial automobility in shaping France's interwar imperial policies and colonial practices.

KEYWORDS   colonial tourism, guidebooks, mobility, technology, business history

In 1923 French author and journalist Jean du Taillis opined that "the tourist today is always a motorist."[1] Having written the first two post–World War I driving guides to Morocco, Algeria, and Tunisia, du Taillis believed that the shift to automobile-based tourism was a foregone conclusion. The emergence of an automotive culture during the interwar years, the democratization of cars, and their commodification within a consumer culture resulted in a new type of leisure travel: automobile-based tourism, or auto-tourism. Yet the spread and development of tourism were not based solely on technological change; tourism's geographic expansion was intimately linked to France's territorial expansion and to its increasing influence and control beyond its continental borders. Where France expanded its colonial presence, French tourists followed. This was especially true

---

1. "Le touriste aujourd'hui est toujours un automobiliste," *Le tourisme automobile en Algérie-Tunisie*, 8.

French Historical Studies • Vol. 48, No. 2 (May 2025) • DOI 10.1215/00161071-11626761
Copyright 2025 by Society for French Historical Studies

for the development of tourism in France's North African colonies, whose geographic proximity to the metropole made them relatively accessible to French tourists. Algeria became an especially sought-after destination, as traveling its three coastal departments, officially parts of France, was essentially a domestic journey for French nationals.[2] Empire thus appeared to function "as a sort of vector along which tourism could grow and expand."[3]

This article explores the relationship between empire and the automobile in the interwar context of French North Africa by focusing on the connection between auto-tourism and colonialism. This type of mobility is seldom examined in tandem with the French imperial project, yet the infrastructure of colonial auto-tourism, as much an industry as a cultural phenomenon and a form of leisure, aptly embodies the complex structure of the empire.[4] Where earlier historical accounts depicted Western technology as a mere "tool" of empire,[5] I draw on the new historiography of automobility to tell a more complicated tale: in the French colonies of North Africa, the interwar rise of automobility generated new relationships between the state, the private sector, and tourists. The resulting dynamic expanded the frontiers of empire and consolidated a more intensive administrative rule. It was nonetheless not exclusively the product of a centralized imperial will.

The burgeoning historiography of automobility,[6] a subfield of the multidisciplinary "mobility turn,"[7] takes movement—of people, objects, and ideas—seriously. A corrective approach to histories that have largely ignored or trivialized movement, new automobility scholarship considers the automobile as more than just a means of transport.[8] Similarly, here the automobile is one element in

---

2. France's empire in North Africa was made up of politically distinctive colonial possessions that enjoyed different legal relationships to the metropole. Algeria's coastal territories, Alger, Oran, and Constantine, were officially integrated into the metropole as departments of France in 1848; the Algerian Southern Territories remained under military control and reported to the governor-general of Algeria; and the protectorates in Tunisia and Morocco were under the administration of the Ministry of Foreign Affairs.

3. Baranowski et al., "Tourism and Empire," 101.

4. Gordon Pirie, who has studied the use of automobiles in colonial Africa, calls for the examination of tourism in connection with the "ebb and flow of technology in imperial times" (Baranowski et al., "Tourism and Empire," 121).

5. Headrick, *Tools of Empire*. For an overview of the scholarship on the history of technology and colonialism, see Arnold, "Europe, Technology, and Colonialism."

6. In its broadest sense, *automobility*, a term combining the words *autonomy* and *mobility*, suggests machines with a capacity for autonomous movement by autonomous persons (Featherstone et al., *Automobilities*, 1, 26).

7. Hannam et al., "Editorial"; Sheller and Urry, "New Mobilities Paradigm"; Urry, *Mobilities*; Flonneau and Guigueno, *De l'histoire des transports à l'histoire de la mobilité?*; Cresswell, "Towards a Politics of Mobility"; Mom et al., *Mobility in History: Themes in Transport*; Mom et al., *Mobility in History: Reviews and Reflections*; Grieco and Urry, *Mobilities*.

8. Urry, "'System' of Automobility"; Featherstone et al., *Automobilities*; Flonneau, *Les cultures du volant*. The existing historiography of automobility remains largely focused on the United States and, in the

a complex whole that comprises users and uses, systems of infrastructure and provision, networks of institutions and organizations, national projects, capital, and a technical object invested with great material power and symbolic force.

The central thrust of this article concerns just how leisure automobile practice was bound up with the imperial project.[9] An examination of colonial auto-tourism complicates the trajectories of interwar *mise en valeur*—the "valorization" of French colonies. It also broadens our understanding of the shift toward automobilization as a means of controlling territories and governing populations. This analytic approach allows for a study of the imperial state that prioritizes technology and materiality; it also shifts the focus to the role of private stakeholders, revealing the role of nonstate actors in the creation and functioning of empire. Finally, this article explores how the automotive imagination intersected with and, at times, amplified the imperial imagination.

Despite the new freedom provided by car travel, auto-tourism did not arise unassisted. The creation of colonial auto-tourism as a modern industry entailed significant efforts in the transportation sector more broadly, creating an infrastructure to carry travelers throughout the empire. It is true that military conquest initially drove transportation and communications infrastructures, spearheading economic penetration and laying the groundwork for interwar colonial auto-tourism. Yet these networks passed from the colonial administration into the hands of manufacturer-entrepreneurs and companies that worked to modernize and rationalize tourism in North Africa. As I show here, in the postwar era commercial enterprises leapt into the business of organizing, promoting, and encouraging colonial auto-tourism. What follows focuses on three such initiatives showing that the construction of colonial automobility was not dictated by politicians and strategy alone; economic agency and commercial motives played a crucial role in creating a new landscape of mobilities. In this process, the car eclipsed (without necessarily replacing) other means of transport, becoming a dominant form of imperial mobility in the interwar era.[10]

European context, Britain. See, e.g., Seiler, *Republic of Drivers*; Sachs, *For Love of the Automobile*; O'Connell, *Car in British Society*; Lavenir, "How the Motor Car Conquered the Road"; and Edensor, "Automobility and National Identity."

9. The scholarship on colonial tourism still largely ignores tourism's essentially material dimensions, mentioning the automobile industry in passing if at all. A recent and outstanding contribution to the historiography of colonialism and automobility is Andrew Denning's *Automotive Empire*, which explores how cars and roads shaped colonial spaces and defined the political, economic, and social dynamics of empire, both within African colonies and in their connections to the European metropole. Another notable exception is Zytnicki and Kazdaghli's *Le tourisme dans l'Empire français*. See also Hunter, "Promoting Empire"; Dulucq, "'Découvrir l'âme africaine'"; Zytnicki, "'Faire l'Algérie agréable'"; Isnart et al., *Fabrique du tourisme*; and Peltre, *Le voyage en Afrique du Nord*.

10. On this point I extend the work of historians who have disrupted the notion of "transition" in studies of energy and mobility by pointing to their nonlinearity and "intensification." Examples of such

The first part of the article contextualizes the automobile in the postwar world of French motorized mobility. The subsequent sections explore three private projects aimed at increasing automobility in the North African empire. The second part looks at the guidebooks developed by Dunlop and Michelin. These books melded product sales, tourism promotion, and a push for new infrastructure with an automotive fantasy of empire that reprised earlier colonial explorations. The third and fourth parts explore the initiatives of businessman John Dal Piaz and industrialist André Citroën. These parts consider Dal Piaz's and Citroën's private initiatives to open up French North Africa by building the technical and commercial infrastructure of tourism. Both men played a key role in building colonial automobility and in tying the North African economy more closely to that of France.

The Rise of Automobility

The automobile was introduced to the African continent well before World War I, and many of the first attempts to establish reliable and orderly colonial automobile routes were in fact reconnaissance missions of a military or quasi-military nature. However, it was only after the war had ended that the push for automobilization of French Africa intensified.[11] By then France's imperial expansion was largely complete. Subsequent years did not see an end to colonial unrest and local insurrection, yet French colonialism shifted from acquiring territories to establishing administrative structures and infrastructures. Among these was a transportation infrastructure, geared to the expanded use of automobiles and considered of prime administrative, political, and military importance.

World War I had a contradictory effect on the automobile industry: while it bolstered the growth of automobile manufacturers, it required the conversion of the automotive industry to military production. This shift into munitions considerably slowed the production of road motor vehicles. Consequently, the number of vehicles in circulation fell dramatically. Nevertheless, the industry prospered. The French automotive industry quadrupled its buildings, machinery, and workforce during the war to meet the demand for military supplies. The use of cars during the war had the unintended consequence of powerfully demonstrating their many practical functions. The iconic vehicles, famously Parisian taxis, requisitioned to transport reinforcements to the First Battle of the Marne in early September 1914 and the vehicles that transported supplies over the *Voie sacrée* to

---

scholarship include Barak, "Three Watersheds in the History of Energy"; Gross and Needham, *New Energies*; and Pinhas, "Moment of Oil."

11. Aix-en-Provence, Archives Nationales d'Outre-mer (hereafter ANOM), "L'automobile à la conquête de l'Afrique (catalogue d'exposition)," 1988.

Verdun in 1916 highlighted the untapped potential of road transport.[12] Though the emergent phase of automobility was interrupted by the outbreak of the war, the war's influence on the development of early automobilism was catalytic rather than interruptive, as Gijs Mom has shown.[13] So it was that the car and the road experienced an undeniable boost during and after the Great War.

The number of automobiles in circulation in France rose dramatically in the years after the war: from 107,535 in 1914 to 750,836 in 1925 and then just over 1.1 million in 1928.[14] A decade later the number of cars had more than doubled, with almost 2.7 million automobiles on the road in 1937.[15] The burgeoning market for cars in the postwar era thus stimulated a considerable output response, with a substantial proportion of total output destined for the export market. The French overseas possessions made for a significant share of the automobile export market outside Europe. Until 1926, 80 percent of French exports were sold on foreign markets outside the French Empire, with 20 percent going to France's overseas territories. However, by 1929, only about 55 percent of exports were directed to foreign markets, while the rest headed to the colonies.[16]

While statistical evidence of automobility in overseas France is sparse for the interwar years, Michelin, France's largest producer of pneumatic tires, provides additional facts and figures: in 1925 there were 9,200 cars in circulation in Morocco, 14,000 in Algeria, and 3,670 in Tunisia. By 1928 those numbers had grown to 17,809, 31,100, and 10,000, respectively.[17] By the end of 1935 the number of automobiles in circulation in Algeria had climbed to 117,117,[18] while a combined 80,000 cars were in circulation in Morocco and Tunisia. With the vast majority of vehicles belonging to the European population in Algeria, there was, on average, one automobile per eight Europeans, a ratio said to be almost double that of metropolitan France.[19] With a total of 256,000 automobiles in circulation in the French colonial empire, the number of cars in North Africa represented 77 percent of all vehicles in the French imperial possessions.[20] Gordon

12. Bardou, *Automobile Revolution*, 79–81; Loubet, *Histoire de l'automobile française*, 65–78.

13. Mom, *Atlantic Automobilism*, 111–13, 227–28.

14. Michelin, *Des faits et des chiffres*, 9.

15. ANOM, "L'automobile aux colonies," 1938, 3 ECOL 27. There is a general paucity of statistics concerning the French automobile industry because it was not included in the *Statistique générale de la France*.

16. Nathan, "L'industrie automobile," 811.

17. The tire maker makes no differentiation between personal and utilitarian vehicles (Michelin, *Des faits et des chiffres*, 10).

18. This includes trucks and light-utility vehicles, buses, and personal cars.

19. Pirie indicates different types of motoring in colonial Africa that mobilized both settlers and visitors, yet the diversity of purpose and experience of colonial automobility prompted a call within the growing global historiography of automobility to develop a distinct critical perspective for settler-colonial automobility. See Pirie, "Non-Urban Motoring"; and Clarsen and Veracini, "Settler Colonial Automobilities."

20. ANOM, "L'automobile aux colonies," 1938, 3 ECOL 27, 12–16.

Pirie argues that, in the colonial context, the make of cars serves to encode patriotism.[21] In this vein, it is significant to observe the considerable presence of American automobiles in interwar French North Africa. Contemporaries spoke of a veritable Franco-American rivalry generated by the penetration of Ford and Chevrolet in the French territories. This was especially true in Morocco, where the French did not benefit from any duty exemptions following the regulation of trade and customs as part of the Act of Algeciras.[22] But for the colonial market to fulfill its potential, French automobile manufacturers worked to protect it from foreign competition by securing a modification of the customs tariffs on automobiles. As Jacques Marseille has shown, a new system of duties, supplemented by strict quotas that replaced the ad valorem tariffs, was applied to Algeria in April 1930. It was extended to Tunisia the following June, ultimately blocking other imports.[23] Data from 1936 indeed reveals that only 954 French automobiles were imported to Morocco versus 1,876 American automobiles, while in Algeria and Tunisia a combined 4,528 French automobiles were imported as opposed to a mere 55 American automobiles.[24]

Automakers as well as tire manufacturers sought colonial markets. The colonial territories served as a preferential outlet and market for the French automotive industry, a rapidly expanding industry of the "future." For Dunlop, Michelin, the Compagnie Générale Transatlantique (CGT), and Citroën, the motorization of the colonies was rich in promise.

Guidebooks to Touring French Africa by Automobile

Narrowly framed, tourist infrastructure refers to liner ships, railways, roads, and hotels: the large-scale technological constructions and the physical forms that enable the movement and accommodation of people. Yet to fully understand the industrialization of tourism, we must focus on infrastructures that operate on various levels simultaneously.[25] As commercial objects with increasingly wide circulation, tourist guidebooks became a key component of the infrastructure of tourism. Far from simple reading, these books operated within a complex assemblage of technical systems, laws and regulations, capital, business interests, and various networks of participants.[26]

---

21. Pirie, "Non-Urban Motoring," 43.

22. All goods imported to Morocco were subject to a 10 percent ad valorem duty plus an additional 2.5 percent tax, regardless of their country of origin (Nathan, "L'industrie automobile," 811).

23. Marseille, *Empire colonial et capitalisme*, 274–75.

24. These numbers represent personal automobiles only (ANOM, "L'automobile aux colonies," 1938, 3 ECOL 27, 19).

25. Larkin, "Politics and Poetics of Infrastructure," 330.

26. This analytic framework draws on the work of science and technology theorists who reject the viability of separating politics and technology into distinct spheres. See, e.g., Latour, *Reassembling the Social*.

Sources of practical information on transportation, accommodation, and places to eat, tourist guidebooks led their readers through space by pointing out "what *ought* to be seen."[27] In the colonial context they played an undoubtedly pivotal role in building an imperial tourist infrastructure: guidebooks mediated between the would-be tourist and the colonial destination. They also promised an experience of "otherness" and "exoticism" that was simultaneously known, familiar, and modern. Intended to be used before, during, and after traveling to the Maghreb, the guidebooks were an all-in-one publication meant to inspire readers to undertake a journey to North Africa.[28] Yet they also appealed to sedentary armchair voyagers, giving them a vicarious travel experience of learning and discovery by bringing the empire "home." In so doing, they built on the cultural infrastructure established by colonial pavilions and expositions. As several scholars have pointed out, colonial tourism extended the tradition of staged colonial representations designed to immerse metropolitan audiences in idealized representations of France's colonies.[29] Like colonial exhibits, guidebooks were structured portrayals meant to contribute to the legibility and comprehensibility of the colonial space.

By the 1920s, there was already a proliferation of tourist guidebooks to North Africa. Among those available to French readers and would-be tourists were Hachette's well-established *Guides bleus*. Also key were the regional guidebooks and travel literature on the Maghreb published by local tourist syndicates (*syndicats d'initiative*) and the nonprofit organization, the Touring-Club de France.[30] The appearance of the first post–World War I touring guidebooks to France's three colonial possessions in North Africa was nonetheless remarkable. *Le tourisme automobile au Maroc* and *Le tourisme automobile en Algérie-Tunisie*, published by the British tire manufacturer Dunlop in 1922 and 1923, respectively, were the first guidebooks designed for automobile-based tourism in North Africa.[31] Important markers of the intersection of auto-tourism and the colonial tourist endeavor, they were joined in 1927 by Michelin's *Maroc, Algérie, Tunisie*. The latter was followed by a second edition in 1930, in celebration of the centennial anniversary of the occupation of Algeria.[32]

27. Koshar, "'What Ought to Be Seen.'"

28. *Le tourisme automobile en Algérie-Tunisie*, 11.

29. Furlough, "*Une Leçon des Choses*," 445–50; Young, "Consumer as National Subject," 282–88. On imperial expositions, see Geppert, *Fleeting Cities*.

30. Established by bicycle enthusiasts in 1890, the Touring-Club de France, the largest nonprofit private tourist association, strived to "develop tourism in all its forms" to facilitate the discovery of the riches of France. Its membership grew rapidly, from nine thousand members in 1894 to seventy-five thousand at the turn of the century, reaching almost one hundred thousand in 1914 (Rauch, *Vacances en France*, 86).

31. *Le tourisme automobile au Maroc*; *Le tourisme automobile en Algérie-Tunisie*.

32. *Guide Michelin: Maroc, Algérie, Tunisie*, 1930. Michelin had promoted French auto-tourism in France's North African empire since before World War I, publishing its *Guide Michelin pour la France et l'Algérie* in 1905 and, two years later, its *Guide Michelin pour l'Algérie et la Tunisie*.

As the closest colonial region, with the strongest and most prolific connections with metropolitan France, the Maghreb was the most obvious location for the expansion of French tourism abroad.[33] Dunlop and Michelin, as well as shipping companies and automobile manufacturers, capitalized on French imperial dominance in expanding their companies' operations into North Africa. Knowing that officials welcomed initiatives that targeted the market for tourism, Dunlop and Michelin consciously cultivated the link between tourism and the colonial endeavor. Their guidebooks offered up a pleasurable recreational experience that celebrated the landed wealth of France's North Africa, opened up the country to French economic interests, and helped legitimize French control. The books also established a vertical order separating users of the automobile from those assumed to be "immobile."

Dunlop's and Michelin's engagement in the Maghreb was also indicative of the sprawling nature of both automobility and tourism, with both businesses seeking a variety of economic opportunities overseas. Michelin already dominated auto-tourism in France with its domestic guidebook series, which introduced regional France to a readership of bourgeois drivers. Expanding the company's commercial initiatives to North Africa meant a new market and an export strategy. Moreover, the majority of European travel occurred during the summer months, while the tourist season in North Africa took place in winter. This seasonal disparity enabled Dunlop and Michelin to generate additional income year-round.

Designed specifically for independent automobile travelers, the Dunlop and Michelin guides helped formulate a new repertoire of travel in French North Africa. In this context, the automobile became central to the consumption of colonial space, with the car lauded as the ideal vehicle for witnessing the marvels of the Maghreb. The guidebooks proclaimed that the road had "dethroned" the railroad and that automobiles had come to "reign without contest."[34] Through such guidebooks, colonial auto-tourism asserted its part in progress and a greater mastery of space; just as important, auto-tourism crystallized the aspirations of bourgeois-aristocratic circles that felt that they had "lost" their independence and individuality due to the limitations of railway travel.[35] The books promoted North Africa as a tourist destination not only for French nationals but also for

33. Elite travel to the Maghreb was common in North African cities as early as the 1850s, with French travelers drawn to the Europeanized littoral (Zytnicki, "'Faire l'Algérie agréable,'" 98–100). Accelerated efforts to create a modern tourist industry in the interwar period started in North Africa, then spread to Indochina and the rest of the empire. For Indochina, see Furlough, "*Une Leçon des Choses*"; and Demay and Herbelin, "La sala et le micro-palace."

34. *Le tourisme automobile au Maroc*, 7.

35. Sachs, *For Love of the Automobile*, 92–97.

Europeans and even North Americans.[36] However, in practice the Dunlop and Michelin guides were designed for a primarily French and overwhelmingly privileged clientele.[37] Unlike earlier manifestations of colonial automobility such as land reconnaissance missions and exploratory expeditions—exclusively male ventures—colonial auto-tourism was an experience promoted as suitable for both men and women.[38] Yet, as a mobile site for the making of multiple subjectivities, colonial auto-tourism was not meant to level social barriers. Nor was it meant to democratize the car or travel more generally. On the contrary, it diverged from the metropolitan trend toward the spread of the automobile among the middle and working classes. In the colonies, auto-tourism reproduced not only classed hierarchy but also the hierarchies of race and culture that bolstered the perceived superiority of modern French civilization.

Self-drive tourists could transport their own vehicles from the metropole or rent an automobile and a chauffeur *sur place*. Dunlop's guides advised against purchasing a new car for the occasion, insisting that *torpédos*—open tourers—would likely suffice. However, should would-be auto-tourists decide to purchase a new automobile, the guides suggested selecting a strong, economical model, comfortable and suited to long-distance travel and uneven road conditions. They also warned against purchasing a *type colonial* with raised axles.[39]

The colonial model of auto-tourism, which combined technological boosterism and the exoticization of Africa, was a crucial element in France's colonial project. By constructing colonial space as an experience to be consumed by car travel, auto-tourism did more than promote the development of colonial transport networks and infrastructure: it also became intertwined with official efforts to expand and consolidate France's North African empire. This is where Michelin's guides stand out: they were the first to combine all three French imperial possessions into one publication, with no differentiation between Morocco, Algeria, and Tunisia, despite their status as distinct polities. The three possessions were themselves governed by two different ministries, yet in combining them the firm aligned itself with colonial ideas of a connective French Africa consolidated by a network

36. In the three decades preceding World War I, the British constituted the second-largest group of European visitors to Algiers. After the war, authorities anticipated an even greater number of British tourists to North Africa following the 1919 nationalist uprising in Egypt (Perkins, "Compagnie Générale Transatlantique," 35, 42).

37. In the 1920s and 1930s French tourists remained people of wealth. Even the 1936 legislation that provided wage raises, a forty-hour work week, and paid vacations (*congés payés*) "did not immediately result in the development of mass tourism or of an expanded commercial tourist industry" (Furlough, "Making Mass Vacations," 258).

38. Images of women travelers appeared in the guides as well as in advertisements for touring North Africa. See, e.g., *Guide Michelin: Maroc, Algérie, Tunisie*, 1930, 44; *L'Afrique du Nord illustrée*, Mar. 22, 1924; and *Les annales coloniales*, July 1925.

39. *Le tourisme automobile au Maroc*, 30; *Le tourisme automobile en Algérie-Tunisie*, 36.

of French transport. In fact, Michelin did not claim neutrality: the preface to the 1927 edition noted that the combined and expanded guide would be a "dependable and knowledgeable mentor" for the predicted wave of visitors from mainland France during the upcoming centenary of the occupation of Algeria. The guide thus directly linked the publication to France's imperial project.[40] The 1930 edition went even further, designating itself the "Centenary Guide" and inviting tourists to witness "French genius" at work in North Africa.[41]

Further, the four guides served as comprehensive introductions to France's North African empire: they provided historical narratives and immersed tourists in the ideologies of empire. The Michelin guides describe North Africa as a region deprived of any "peace, prosperity, or civilization" prior to the arrival of the French, who "renewed and expanded the tradition of Rome." Had the people of North Africa been "left to their own devices," the books suggested, they would have known complete disorder.[42] Likewise, the Dunlop guides laud "France's colonizing genius" and deny the region any Indigenous civilization.[43] Disorder reigned, they claimed, until the arrival of the French, but through sound French administration the region had experienced a true "renaissance."[44] As Diana K. Davis has explained, a carefully constructed narrative that "the French were the legitimate heirs of Rome in all her imperial glory" was used to justify French occupation and expansion across North Africa. This narrative helped promote the colonial project and sat well with the imperial goals of "improvement." As the true successor of Rome, France had a duty to revive the Roman oeuvre and to "restore" North Africa to its mythical former fertility. Integrated into an imperial imaginary, this idea had very real social, political, and economic implications.[45]

The guidebooks perpetuated the notion that the arrival of the French heralded the so-called rational exploitation of natural resources, a rhetoric that drew on a colonial vision of Africa as a continent awaiting possession and mastery. Indeed, the guides underscored that it was only due to France's industrious administration and modernizing work that motorists could now take advantage of a French-built North African road network. "Lucky" motorists were invited to enjoy a modern system of roads "comparable to that of France."[46] The provision of a modern infrastructure encapsulated French ingenuity and rationality and was represented to readers as a wondrous achievement, a "work of art" even,

40. *Guide Michelin: Maroc, Algérie, Tunisie*, 1927, ii, foreword.
41. *Guide Michelin: Maroc, Algérie, Tunisie*, 1930.
42. *Guide Michelin: Maroc, Algérie, Tunisie*, 1927, xi; *Guide Michelin: Maroc, Algérie, Tunisie*, 1930, 30.
43. *Le tourisme automobile au Maroc*, 9, 17, 20; *Le tourisme automobile en Algérie-Tunisie*, 9.
44. *Le tourisme automobile en Algérie-Tunisie*, 25.
45. Davis, "Restoring Roman Nature."
46. *Le tourisme automobile en Algérie-Tunisie*, 7; *Guide Michelin: Maroc, Algérie, Tunisie*, 1930, 50.

due to its "modern techniques" and "state of the art maintenance."[47] This superior French modernity was juxtaposed with the supposed ignorance of local industries toward newer technologies like steam and electricity.[48] Only on rare occasions did the guides depart from narratives of French progress, generally to promote their own business interests. The 1922 Dunlop guide to Morocco, for instance, maligned French-constructed roads in the colonies to promote sales of its own tires: "It is quite certain that the roads cannot be as safe as those of the *mère patrie*. . . . So, if you have a good car . . . it must also be fitted with good tires!"[49]

At the same time, touring the colonies put the French in direct contact with France's imperial project. Historian Ellen Furlough has shown that France promoted colonial tourism in a propagandistic effort to foster the patriotic identification of French citizens with "their" empire. Touring the colonies was a pedagogical means of "knowing" the empire by providing the tourist with firsthand "facts," however selective and limited.[50] Du Taillis, author of the two Dunlop guides, invited his readers to "travel, admire, enjoy!" He also suggested that travelers who undertook the journey had "understood *their* duty [having] finally decided to travel *their* opulent provinces of *our* North Africa" (emphasis added).[51] Du Taillis was a writer for *Le matin* and *Le Figaro* as well as a correspondent for the *Oran-Matin* in Algiers. Through his guides for Dunlop, tourists not only got to experience the empire in the Maghreb and see its achievements firsthand but were also a part of the ongoing French national-imperial project overseas. Traveling the classic tourist routes thus served an "instructive and reinvigorating" role, revealing to motorists the "persistent labor of *our* settlers" (emphasis added).[52] The itineraries highlighted the patrimonial importance of places visited and paid homage to France's colonial heroes and its conquest of North Africa. For example, motorists who traveled Kabylia would follow routes "painfully" forged by "*our* soldiers"; en route to Fort-National (Larbaâ Nath Irathen) they would follow "a magnificent road traced in 27 days by *our* soldiers when conquering this wild and tormented land" (emphasis added). Additional directions took motorists through the Zaatcha palm groves, known for the "murderous siege that *our* troops sustained for 52 days during the 1849 insurrection," and the journey to Touggourt took them through Megarine, the "last and bloody" site of the colonist who planted the *tricolore* on the city's mosque

47. *Le tourisme automobile en Algérie-Tunisie*, 7; *Guide Michelin: Maroc, Algérie, Tunisie*, 1930, 31.
48. *Le tourisme automobile au Maroc*, 23.
49. *Le tourisme automobile au Maroc*, 31.
50. Furlough, "*Une Leçon des Choses.*"
51. "*Circulez, admirez, profitez!,*" *Le tourisme automobile en Algérie-Tunisie*, 8.
52. *Le tourisme automobile en Algérie-Tunisie*, 84.

(emphasis added).[53] Dunlop's guide not only promoted collective identification with the empire through the new possibilities of automobility but also formulated an imperial tradition of French rule in North Africa wherein motorists were the natural successors to mythologized explorers, adventurers, colonists, and officials. By representing certain itineraries as "motorized reenactments" of earlier colonial conquests, now performed within "pacified" spaces, the guides invited tourists into the tradition of colonial expansion. The act of driving became an important practice of colonial occupation.[54]

Now that trips were no longer constrained by the existence of train lines, Dunlop and Michelin had to organize their guidebooks according to drivable itineraries. Each designed to fuel desires for colonial travel by introducing France's North African possessions to a metropolitan readership, the two sets of guidebooks sought to navigate tourists through colonial space differently. The Dunlop guide to Morocco comprised 194 pages and included seven color maps, thirteen city plans, sixteen itineraries, and 100 photographs, while the guide to Algeria and Tunisia, published just a year later, boasted almost double the number of pages, with seven color maps, thirty-four city plans, thirty itineraries, and 160 photographs. The Michelin guidebooks featured a considerably higher number of possible tours divided into two types: eighty-one *grands itinéraires directs pour automobiles* (longer routes by car) followed by twenty-six *excursions* (shorter trips). The Michelin guides also differed in their cartographic coverage: the company's guides provided black-and-white maps for forty-three communities, and seven color maps for the larger cities of Algiers, Casablanca, Constantine, Fez, Marrakesh, Oran, and Tunis. They also included simple maps of the twenty-six excursions and offered a regional map of Morocco, Algeria, and Tunisia, an "indispensable complement" to the guidebooks. The map was available for purchase for 6 francs for the paper version and 16 francs for the cloth version. The firm's extensive mapping was the product of André Michelin's experience as a cartographer at the Ministry of the Interior coupled with his understanding of what would be useful for tourists. At first Michelin had depended on official *Etat-Major* cartography commissioned for military uses, as well as on maps drawn up by the Ministries of Public Works and of the Interior for maintenance and construction purposes. Over time, the impracticability of these maps for motorists drove the manufacturer to create its own maps tailored to nascent automobile tourism.[55]

While both Michelin and Dunlop set out to make colonial space more knowable, they also helped define what was of touristic interest. Meant to pique

53. *Le tourisme automobile en Algérie-Tunisie*, 159, 162, 214, 218.
54. In these cases, colonial automobile-based tourism partly aligns with the paradigm of settler-colonial automobility as outlined in Clarsen and Veracini, "Settler Colonial Automobilities," 894.
55. On Michelin's role in mapping France, see Harp, *Marketing Michelin*, 74–79. See also Olson, "Come Drive French North Africa," 33.

drivers' interest and fuel their desires for colonial auto-tourism, both companies maintained the image of the colonies as exotic, wild, and enchanting locales. The regions penetrated and traversed are dotted with active, fertile, thriving, and prosperous centers of colonization that testify to France's colonial impetus.[56] Meanwhile, the books rendered Indigenous peoples and places as entertainment, in the long imperial tradition of spectacle and exoticization.[57]

Driving directions are thus superimposed on observations of the French colonial project. The guidebooks reproduce hierarchical classifications of both places and people, reinforcing the distance between metropolitan and native peoples, cultures, and landscapes. At the same time, they seek to shrink the distance between metropole and colonial periphery, enhancing the idea and vision of *la plus grande France*.[58] The Dunlop and Michelin guides were especially useful in accommodating both the unity and diversity of Greater France, contrasts that became only more stark in the era of automobility: "It has been said and repeated of North Africa that it is the country of contrasts . . . where two cultures coexist but seem to have remained foreign to one another. . . . On the road next to a small donkey . . . an enormous and luxurious bus. . . . In the desert, on the same track, a turbulent and colorful caravan meets that of six-wheeled cars or *autochenilles*."[59] A distinctly imperial narrative, wherein powerful means of transport are substituted for "obsolete" ones such as camels and donkeys, this description weaves automobiles into a narrative of linear progress.[60] Doing so in turn highlighted French dynamism and achievements: the six-wheeled Renault automobiles and Citroën's *autochenilles* were the vehicles used by the *grands constructeurs* on their first trans-Saharan expeditions.[61] These technically modified and improved motor vehicles were developed to enable movement across various terrains, including the Sahara. The *six-roues* (six-wheel) Renaults were light vehicles, equipped with a drive shaft connected to two rear axles fitted with twinned wheels. The axles were designed so that they could each move vertically

---

56. *Le tourisme automobile en Algérie-Tunisie*, 84, 88, 89, 129, 130, 281.

57. *Le tourisme automobile en Algérie-Tunisie*, 55, 163, 218, 257, 304.

58. Furlough, "*Une Leçon des Choses*," 444. The concept and figure of Greater France were more fervently debated and imagined during the interwar period. See Wilder, "Framing Greater France Between the Wars." Joseph Bohling has also examined French expansion strategies in the interwar period through a larger European political entity, a "continental bloc," rather than through the French "empire-state" ("Colonial or Continental Power?").

59. *Guide Michelin: Maroc, Algérie, Tunisie*, 1930, 44.

60. The landscape of mobility depicted in the guides, dominated by the automobile, describes a linear path toward modernity in which motorized traffic replaced nonmotorized traffic. Yet, in their descriptions and itineraries, the guides also reveal that new transport technologies did not necessarily replace traditional and nonmotorized modes of mobility.

61. The 1922–23 Citroën desert crossing was first, followed by two Renault-sponsored expeditions in 1924, and then Citroën's 1924–25 expedition across the Sahara.

on their own, changing their position according to changes in the terrain. The *autochenilles* had wheels in the front and a patented caterpillar halftrack propulsion unit in the back, developed by the engineer Adolphe Kégresse. The new system had a flexible belt of canvas and rubber placed around a drive wheel at one end, an idler wheel at the other, and several guide wheels in between, expanding the lifting surface of the vehicle and enabling it to increase traction and speed. The Saharan crossings were a pivotal moment in France's aggressive pursuit of a motorized and mobile colonial Africa in the immediate postwar years; they are an exemplar of what Andrew Denning has called the "techno-mobile" logic and vision of empire that became paradigmatic in the French colonialism of the interwar era.[62] Moreover, the trans-Saharan expeditions were harbingers of subsequent colonial conquest by car.

In the circuits they suggested, and in their descriptions of itineraries, Dunlop's and Michelin's guidebooks for French North Africa navigated motorists through regions that offered "a particular tourist interest."[63] The tire makers played a seminal role in creating the new automobility that quite literally drove the territorial spread of France's empire in North Africa. Their guidebooks promised an experience far removed from everyday metropolitan life, offering encounters with strikingly different environments and peoples. The Dunlop and Michelin guides not only equipped motorists with the information needed to make the "inaccessible" independently accessible by car but also trained them in imperial ideologies and promoted an imperial consciousness.

Auto-Tourism and the Compagnie Générale Transatlantique

Dunlop and Michelin set out to guide an overwhelmingly French contingent of self-drive tourists through North Africa. This, however, was not the only option. Organized automobile tours were another driver of modern tourism infrastructure in the Maghreb, one that opened a wider range of opportunities to an even wider circle of social elites.

One of the most influential promoters of auto-tourism to the Maghreb was John Dal Piaz, who was named president of the CGT in 1920.[64] The company initially specialized in shipping, but under his presidency it launched the Auto-Circuits Nord-Africains, a business venture that endeavored to transform, modernize, and standardize transport and travel in French North Africa.[65] To

---

62. Denning, "Mobilizing Empire," 66.
63. *Guide Michelin: Maroc, Algérie, Tunisie*, 1930, 155.
64. The CGT was founded in 1855, first operating maritime routes in the Atlantic. In 1879 it began operating in the Mediterranean. Its first crossing between Marseille and Algiers took place on June 2, 1880.
65. From 1925 the company was known as the Société Voyages et Hôtels Nord-Africains.

that end, the Auto-Circuits set out to create a tourist-oriented infrastructure of transport, hotels, tours, and itineraries. This shift diversified the company's function, from primarily serving industrialization and hauling mail and freight to facilitating travel for the purpose of leisure. The CGT envisioned an all-inclusive travel package: tourists would sail to Africa aboard the company's liners, stay in company-built hotels, and travel across the Maghreb in company-organized and chauffeur-driven automobile tours. The company further developed and operationalized the concept of package tours, an innovative concept that transformed the experience of travel.[66] Creating a total and coordinated experience from start to finish was essential in convincing potential tourists of the viability of touring the Maghreb.

Tasked with rebuilding the CGT after the war, Dal Piaz believed that the postwar years were a propitious moment for increasing access to France's North African possessions. The CGT would do so by creating a modern, tourist-oriented infrastructure of automobile travel services, first from Morocco to west Algeria, then across Algeria, and finally to Tunisia and into the Southern Territories—all accompanied by major publicity efforts. In its first year of operations, the company administered four automobiles that transported a total of thirty tourists over 110,000 kilometers. During the 1925–26 season the numbers had risen to 195 company cars and 790,000 kilometers traveled. In 1928, the company boasted 280 automobiles, moved roughly 5,000 passengers, and had completed 1,450,000 kilometers of touristic itineraries.[67] By 1929, a decade after the Auto-Circuits began its operations, over forty thousand tourists had "already appreciated its incomparable facilities."[68]

The CGT's all-inclusive, single-price travel package was lauded as a practical and effective means of introducing tourists to France's North African possessions. The package included sea crossing, city-to-city train journeys, luggage transfer, automobile excursions, local guides, hotel stays, food (but not drinks), and tips. Such trips organized consumption of France's colonies in the most convenient and secure way, insulating tourists from the burdens of uncertainty. Even in its efforts to promote individual auto-tourism, the Dunlop guides lauded the CGT's travel packages. The company's guide to Morocco described the CGT as an "audacious enterprise" that truly understood the meaning of *la mise en valeur*, turning North Africa into a bustling tourist hub and a "new source of wealth and propaganda" for France's colonial domain.[69] Dunlop's

---

66. This model of tourism offered by professional organizers who arranged fixed itineraries and handled routes and accommodation was developed by Thomas Cook more than any other promoter and organizer of leisure travel. See Hunter, "Tourism and Empire."

67. Ricard, *Le grand tourisme*, 14–16.

68. Vox, *Les auto-circuits nord-africains*, 63.

69. *Le tourisme automobile au Maroc*, 35.

guide to Algeria and Tunisia celebrated the CGT's initiative in the Maghreb for "serving so deeply the cause of tourism in North Africa." It went further, declaring that the company "deserves much more than the acknowledgments of travelers. . . . [It deserves] the gratitude of France for having contributed so powerfully to spreading its colonizing virtues of great renown."[70]

The CGT's colonial excursions catered to the growing market for tourism to the Maghreb among elites—well-to-do tourists who had the time, leisure, and means for travel. For their Maghreb adventure, the company's affluent clientele could choose between a group trip in a company-owned vehicle with ten armchair seats and a private trip that accommodated up to five passengers. The company cars, or *autos TRANSAT*, were advertised as automobiles that met the multiple needs of North African tourism: they were robust, fast, and comfortable. Most important, they "make it possible to visit the most picturesque sites and to reach regions until now ignored by tourism."[71]

As early as April 1920, the CGT offered a three-week package tour beginning either in Marseille or in Bordeaux. From Marseille tourists would sail to Algiers and then embark on a seventeen-hundred-kilometer itinerary to Casablanca via Ténès, Oran, Fez, and Rabat. The Bordeaux trip followed this same itinerary in reverse. The tour cost between 4,950 and 6,500 francs per person for an all-inclusive *voyage collectif*, while the cost for a *voyage particulier* was anywhere between 5,050 and 15,500 francs per person, depending on the type of car and the number of people in the group.[72] By the 1929–30 season those prices had soared to 9,000 or 10,600 francs per person traveling in a group or anywhere between 9,600 and 25,500 per person for a *voyage particulier*.[73]

In October 1920 the CGT held the inaugural Algiers–Casablanca–Marrakech voyage. Members of the CGT were joined by important French government personnel, local colonial administrators, and representatives of the Touring-Club de France and the Automobile-Club de France.[74] Also aboard were French and foreign members of the press, who had been invited to the debut of the "great tourist route" from Algeria to Morocco.[75] The *caravane transatlantique* garnered

70. *Le tourisme automobile en Algérie-Tunisie*, 31–32.

71. Tourists who traveled in their own vehicles incurred an additional charge for the transportation of an automobile and driver from the metropole to Africa (Vox, *Les auto-circuits nord-africains*, 63).

72. Le Havre, French Lines et Compagnies, *Auto-circuits nord-africains de la Compagnie Générale Transatlantique (dépliant informatif)*, n.d., 1997 004 7423.

73. Costs also represented seasonal price variations from October to January and from January to May (Vox, *Les auto-circuits nord-africains*, 9).

74. The Automobile-Club de France was established in 1895 and boasted a smaller and more exclusive membership than the Touring-Club de France. Both André and Edouard Michelin were founding members of the Automobile-Club de France (Harp, *Marketing Michelin*, 54–56).

75. Among the distinguished invitees were M. Norrel, undersecretary of State for Hydraulics; M. Robert David, undersecretary for the Ministry of the Interior; M. de Beaumarchais, director of the Africa

the official and enthusiastic support of local government officials, including the governor-general of Algeria, Jean-Baptiste Abel, and the resident-general of France in Morocco, Hubert Lyautey, who held receptions in honor of the achievement.[76] Such support within the high colonial administration demonstrates that tourism and transportation infrastructure between the colonies were part and parcel with colonial development at large.

In March 1921 the CGT announced two new itineraries, from Algiers to Tunisia and from Algiers to Laghouat, an oasis on the north edge of the Sahara Desert. By November 1923 the CGT offered ten touristic circuits, and by the 1929–30 season twenty-five. The itineraries ranged in length and duration. The shortest itinerary, from Algiers to Bou Saada, situated about 245 kilometers south of the coastline, traversed Kabylia, a mountainous coastal region in northern Algeria. The entire route was 872 kilometers long and was to be completed over eleven days. The cost ranged from 4,100 francs per person in a group trip to 11,900 francs per person for a private tour.[77] Unlike earlier CGT itineraries that traversed the Maghreb along the coast, the longest itinerary, the Grand Erg tour, was a 3,221-kilometer circuit that ventured far inland. Starting in Algiers, it made a thirty-five-day round-trip voyage through the oases of El Golea (El Menia), Ghardaia, Laghouat, and Bou Saada. Accessible only by "special Saharan cars"—that is, by the specially equipped Renault automobiles mentioned earlier— the cost of travel for this circuit was between 17,500 francs for a group tour and 34,000 francs for a private tour of up to four persons.[78] As was true of other tours to the Saharan oases, including Ouargla and Touggourt, the Grand Erg tour enabled the wealthiest and most adventure-loving tourists to travel far into the continent and its desert regions. In fact, Timimoun, the tour's southernmost point, was located fifteen hundred kilometers from the Algerian coast. Tourism thus pushed the frontiers of empire far into the Algerian desert.

Driving the process was the powerful colonial and economic urge to solve the problem of locomotion across the Sahara. The latter was firmly rooted in the French imagination as an exceptional space and as the ultimate challenge to colonial conquest. Fantasies of conquering the Sahara had existed decades prior to the outbreak of World War I.[79] Only after the war did entrepreneurs, French

---

Division of the Ministry of Foreign Affairs; M. Rondet-Saint, president of the French Maritime League; and M. Famechon, president of the National Tourism Office (*L'écho d'Alger*, "Tourisme").

76. *France-Maroc*, "La route touristique Alger-Casablanca-Marrakech"; *L'écho d'Alger*, "Le circuit Nord-Africain."

77. Vox, *Les auto-circuits nord-africains*, 25.

78. All prices were per person (Vox, *Les auto-circuits nord-africains*, 61).

79. The most ambitious colonial technological solution to the challenge of mobility and transport across the desert was the trans-Saharan railway, yet this scheme never materialized (Heffernan, "Shifting Sands").

administrators, and military men form a powerful alliance around the shared goal of Saharan development. Arnaud Berthonnet has explored the key role of the military in this process: due to their expertise in land reconnaissance in the French African colonies, military personnel played a central role in advancing and organizing mechanized movement in the Sahara, including tourism. They surveyed the landscape and marked out *pistes* and fuel and oil depots in the desert, participated in trans-Saharan expeditions, founded trans-Saharan automobile transport companies, and published tourist guides to the great desert.[80] Moreover, unlike Alger, Oran, and Constantine, the Southern Territories, which spanned the entire Algerian Sahara, remained under military administration and were not considered part of metropolitan France. Consequently, travel in the Sahara was a doubly militarized experience: the military played a key role in opening the Sahara to tourists, and while there, travelers depended for their security on the French Army, charged with maintaining and administering order in the desert. The emphasis on the impenetrability of the Sahara Desert, now reputed to be "open" and "accessible" by virtue of technological modernity, was a distinctively imperial narrative. At best this narrative ignored the travel networks that had developed and flourished across the Muslim world for many centuries prior. In actual fact, imperial travel in the Sahara interrupted or repurposed these preexisting networks. Nevertheless, in the minds of French administrators, it was the car that finally tamed the Sahara.[81]

Dal Piaz capitalized on contemporary changes in mobility to promote the automobile as a consumer product and to create new opportunities for colonial automobility. As a result, he was fully involved in delineating and broadcasting the French presence. The infrastructure organized and mechanized by the CGT served the needs of the growing number of people who desired to tour and sojourn in the French colonies. However, this infrastructure for leisure travel was also central to the broader patterns of mechanized colonial mobility that helped territorialize French imperial power. Private initiatives like those of the CGT "simply" sought to "realize transport enterprises in France, the French colonies, and

80. Berthonnet, "Le rôle des militaires français." Some noteworthy examples include the aviation officer Lieutenant Louis Audouin-Dubreuil, who helmed Citroën's first and second trans-Saharan expeditions alongside Georges-Marie Haardt; General Octave Meynier, director of the Territoires du Sud, who compiled a *Guide pratique du tourisme au Sahara* in 1931 with Captain Albert Nabal; and the Compagnie Générale Transsaharienne, founded by the entrepreneur Gaston Gradis and Louis Renault, in which General Jean-Baptiste Estienne and his son, the military pilot Georges Estienne, served as chief executive officer and general manager, respectively. Retired military officers remained involved in the automobilization of the Sahara after World War II, notably through the Rallye Méditerranée–Le Cap, a car race from Algiers to Cape Town held four times during the 1950s and once more in 1961. See Brown, "Le Rallye Méditerranée–Le Cap."

81. Various studies have reevaluated the role of local knowledge and preexisting networks and patterns of movement and exchange in colonial technological modernity. See, e.g., Chatty, *Nomadic Societies in the Middle East and North Africa*; Fletcher, "Running the Corridor"; and Pétriat, "Uneven Age of Speed."

countries under French protectorate, mainly in North Africa."[82] In doing so, however, they became essential to the construction, development, and expansion of colonial mobility during the interwar period.

André Citroën and Colonial Tourism

Another primary promoter of colonial travel and tourism in the first half of the 1920s was the French businessman-industrialist André Citroën, founder of the eponymous automaker. Motivated by the success of his first African venture, crossing the Sahara by automobile in 1922–23, Citroën turned to developing commercial automobile traffic in the colonies. For him, that first expedition demonstrated that automobile travel across France's colonial possessions was both possible and practical. As he put it, "One of the immediate consequences to be drawn from the *raid africain* . . . is grand tourism; all that is needed is to organize supply lines and mark out the itineraries."[83]

As a result, Citroën began planning a "vast program" for Africa.[84] The second step was an African transport company, the Compagnie Générale Transafricaine (CEGETAF), later renamed the Compagnie Transafricaine Citroën (CITRACIT), as well as a second trans-Saharan expedition, best known as the Central Africa Expedition (Expédition Citroën Centre-Afrique) or the Croisière Noire. The planned African transport company was first pitched in January 1924 and scheduled to be inaugurated a year later, in January 1925. The company would have officially initiated Citroën's grand scheme to develop regular automobile services in all of France's African territories, but it was canceled prior to the start date. Despite never having materialized, CITRACIT offers a revealing case study.[85] Official and enthusiastic support for Citroën's CITRACIT demonstrates the political stakes of automobiles in the material transformation of French Africa. It also demonstrates the role of the metropolitan imagination, which envisioned an expanded African infrastructural network as the basis for imperial development, even when at stark odds with the realities on the ground. This dual attempt to control French colonial space, both geographically and symbolically,

82. French Lines et Compagnies, *Statuts de la Société des voyages et hôtels nord-africains: Auto-circuits nord-africains*, 1925, 1997 004 7422.

83. Hérimoncourt, Centre d'Archives de Terre Blanche (hereafter CATB), La Compagnie Transafricaine Citroën, n.d., DOS2016ECR-01057, 9.

84. Paris, Fonds Georges-Marie Haardt, Musée du Quai Branly (hereafter FGMH/MQB), "Projet d'organisation des grandes lignes de communication africaines," n.d., DA001110/64069, 9.

85. Citroën sought to withhold information and stifle any discussion of the project, thus "actively erasing" the existence of the CITRACIT, which has consequently slipped out of historians' views. An exception is Alison Murray's research on the industrialist's private colonial tourism enterprise in Africa, again, with a particular focus on film and its use to promote and disseminate colonial imagery ("Le tourisme Citroën"; "Les automobiles, le désert africain, et le cinéma").

is an example of the workings of the new colonial policies of the interwar period and of the entanglement of empire, industry, and technology.

Various studies have explored Citroën's second trans-Saharan expedition. Some include informative albeit fragmentary overviews of the crossing, while others focus on recapitulating the expedition by echoing its official narrative as it appears in the travelogue written by expedition leaders Georges-Marie Haardt and Louis Audouin-Dubreuil.[86] Notably, Peter J. Bloom explores the expedition through the lens of Citroën-sponsored interwar films on the trans-Saharan crossing and as part of the metropolitan culture of popular imperialism.[87] Here I contextualize Citroën's first expedition as part of his grand scheme to create an extensive network of automobile routes across the continent. This was a program that included a colonial auto-tourism initiative, an integral yet overlooked part of Citroën's *longue durée* vision of an "empire on wheels." Here I follow Andrew Denning's study of the Citroën Central African Expedition, which he portrays as a moment when French colonial authorities turned to transportation and mobility as they pursued the economic and political goals of empire. Denning's analysis "reveals a complex public-private partnership that complicates understandings of centralized, state-driven colonial initiatives in interwar French Africa."[88]

Citroën believed that the development of a vast transportation infrastructure should be the principal outcome of his first trans-Saharan expedition. In particular, he envisioned the creation of extensive networks of automobile routes across the continent. For Citroën, the creation of first medial and then lateral grand arteries of communication "with the help of automobiles" and "using modern technology" would connect the French colonies following a rational method of development. Each path of communication would first be surveyed, then developed, and finally exploited as an operational network of automobile routes.[89] His vision was that of an African continent crisscrossed by long-distance routes and serviced by technologically advanced transport systems. The network was to include five transcontinental lines of transport, from the Mediterranean to the Gulf of Guinea, from the Atlantic Ocean to the Gulf of Aden, from Chad to Congo, from the Mediterranean to Chad, and from the Atlantic Ocean

86. Sabatès, *La croisière noire Citroën*; Bejui and Bejui, *Exploits, fantasmes transsahariens*; Audouin-Dubreuil et al., *Les croisières Citroën*; Audouin-Dubreuil, *La croisière noire*.

87. Bloom, *French Colonial Documentary*; see also Murray, "Film and Colonial Memory."

88. Denning, "Mobilizing Empire," 47. In his examination of the business relationship between the Beirut-based Ford dealer Charles Corm and political-administrative functionaries in French Syria, Simon Jackson claims that interpersonal relationships quite literally facilitated movement and mobility in the interwar French Levant ("Personal Connections and Regional Networks").

89. FGMH/MQB, "Projet d'organisation des grandes lignes de communication africaines," 2–3, 7.

to the Indian Ocean. The network's itineraries, divergent lines, and possible extensions encompassed an incredibly ambitious automobile communications complex extending over 22,500 kilometers.[90]

Citroën corresponded with French administrators in different branches of the government, from the president of the council to the ministers of colonies, war, and foreign affairs. In his missives, he articulated his commercial initiatives in a language that brimmed with references to nation and empire. The entrepreneur offered a host of benefits to the Republic and its colonial apparatus as he sought to mobilize official support for his African endeavors. He brandished the national and colonial contributions of his projects and their numerous political, economic, and military advantages and gains. Moreover, he maintained that cars and roads would "weld together" French colonies in Africa into a coherent geopolitical unit—the desired French African Bloc—rather than a fragmented territory.[91]

Citroën was determined to change colonial practices of movement by creating networks of mechanized mobility. In planning the networks that would connect colonies to one another, Citroën also believed it necessary to plan and promote touristic movement in regions favorable to "our overseas nationals."[92] Initially, the CEGETAF was to offer a twenty-six-day round-trip tour from Colomb-Béchar in Algeria to the Niger in *autochenilles*. The service was meant to depart every four days, between November and the beginning of March.[93] It was an all-inclusive package, consisting of visits to Oued Saoura, Timoudi, Ouallen, Tessalit, and Bourem, among other locales deemed of touristic interest. In Gao, for example, the caravan would split, offering either a touristic circuit in Timbuktu or a hunting safari in the W region, named after the W-shaped meander of the Niger River, located near Niamey and known for its abundant game. Tourists would sojourn in five company-owned *bordjs* (forts) that served as hotels, located in Colomb-Béchar, Béni-Abbès, Adrar, Gao, and Timbuktu, as well as seven comfortable company-run camping areas. Citroën emphasized the modernity and comfort of his initiative, drawing attention to the company's construction

90. FGMH/MQB, "Projet d'organisation des grandes lignes de communication africaines," 5–6.

91. FGMH/MQB, "Lettre d'André Citroën au ministre des Colonies," Apr. 30, 1924, DA001110/64069. Citroën reiterated his goal of fusing French colonies in North, West, and Equatorial Africa into a continental bloc in letters to other government officials, including the president of the council, the minister of foreign affairs, and the minister of war. See also Denning, "Mobilizing Empire," 54, 56.

92. FGMH/MQB, "Organisation d'une voie de communications régulières par autochenilles entre l'Afrique du Nord et l'Afrique Occidentale française," n.d., DA001110/64069, 1–2.

93. In later presentations of the tourist initiative to French administrators, Citroën changed the frequency of departures to every eight days, and eventually to every fifteen days (FGMH/MQB, "Lettre de la direction générale usines Citroën au président du Conseil, ministre des Affaires étrangères," May 2, 1924, DA001110/64069).

of commodious lodgings equipped with electricity and running water.[94] In addition to the Colomb-Béchar–Timbuktu circuit, the CEGETAF would also offer automobile excursions to the mountainous region of the Hoggar, as well as shorter trips from Colomb-Béchar to the Saharan oasis of Touggourt during the month of March, before the hot season.[95]

In its construction phase, CITRACIT became known for its efficacy and prompt operation. According to Citroën, the construction of four of the five planned *bordjs* was completed in six months by a crew of 150 men, equipped with twenty *autochenilles*.[96] He prided himself on the extraordinary rapidity with which he was able to execute his project and believed that he had once again proved the rationality of his endeavors and his superior inventiveness:

> The news of the project was just made known, and already comfortable *bordjs* sprouted from the ground in the middle of the desert. . . . Cars equipped with *chenilles* will be able to transport everything through the chaos of the desert lands, while other vehicles modified for their new destination . . . designed, built and shipped . . . [transport] the first travelers from one bank of the Saharan sea to the other.[97]

In anticipation of inauguration day, various advertisements and publicity brochures marketed the CITRACIT's excursions, including an offer for a twelve-day, twenty-two-hundred-kilometer trip from Paris to Timbuktu. The biweekly service would include transfer by rail from Paris to Marseille, sea crossing from Marseille to Algiers, and a thirty-two-hour train trip to Colomb-Béchar, the starting point of CITRACIT's desert itinerary. The adventurous Saharan excursion from Colomb-Béchar, traversing the heart of French Sudan to Timbuktu by automobile and *autochenille*, would take eight days. The shortest travel day traversed 170 kilometers, from Béni-Abbès to Timoudi, and the longest over 510 kilometers, from Gao to Timbuktu.[98] In Gao passengers would have the option of either continuing by automobile to Niamey for a safari hunt or transferring

94. FGMH/MQB, "Organisation d'une voie de communications régulières par autochenilles entre l'Afrique du Nord et l'Afrique Occidentale française," 2–5.

95. FGMH/MQB, "Organisation d'une voie de communications régulières par autochenilles entre l'Afrique du Nord et l'Afrique Occidentale française," 3–6; FGMH/MQB, "Lettre de la direction générale usines Citroën au président du Conseil, ministre des Affaires étrangères"; FGMH/MQB, "Lettre d'André Citroën au ministre des Colonies."

96. CATB, *La Compagnie Transafricaine Citroën (brochure illustré)*, n.d., PI2015ECR-01657. In her research Murray found photos and maps confirming the construction of *bordjs* in Colomb-Béchar, Béni-Abbès, and Adrar. However, she found no concrete evidence of the existence of the Gao and Timbuktu *bordjs* ("Le tourisme Citroën," 101).

97. FGMH/MQB, "Lettre de la direction générale usines Citroën au président du Conseil, ministre des Affaires étrangères."

98. The complete itinerary was Colomb-Béchar, Beni-Abbès, Timoudi, Adrar, Ouallen, Tessalit, Tabankort, and Gao.

to river transport and traveling along the Niger to the port of Timbuktu. The company's service, which included transportation, food, and board, was affordable only for the very well off: a round trip to Timbuktu cost 40,000 francs; an additional three-day trip to Niamey cost 5,000 francs, while an additional fifteen-day hunting excursion cost 15,000 francs (not including weapons and ammunition).[99]

CITRACIT's inaugural run was planned for January 1925, and the invitees included distinguished passengers such as the governor-general of Algeria, Théodore Steeg; Maréchal Philippe Pétain; Albert Sarraut; M. Boulogne, director of the Southern Territories; M. Rouzaud, director of the Algerian State Railways; and King Albert I of Belgium.[100] The inauguration was postponed following a report sent by Colonel Dinaux, military commander of Aïn-Séfra in the Southern Territories, to Governor-General Steeg, on tribal unrest on Morocco's southern border. The report included additional information on rebel groups preparing to attack automobile convoys.[101] Alison Murray suggests that Dinaux may have exaggerated the danger because he himself was hostile to the prospect of an influx of tourists traversing the Sahara in automobiles.[102] Regardless of motivation, Dinaux's reports gave rise to serious questions about the possibility of completing the voyage safely. Consequently, King Albert I withdrew his participation. *Le matin* published a release issued by the Belgian cabinet, which cited scheduling concerns as the reason for the king's cancellation. It reported that, following the cancellation and the apparent danger and difficulty in assuring the safety of the convoy and its passengers, Citroën had temporarily postponed the inauguration of the Colomb-Béchar–Timbuktu line until these "unforeseen circumstances" had passed.[103] Less than a week later, newspapers reported that Citroën had officially abandoned his project due to the menacing activity of a group of dissidents from Tafilalet, which had ambushed colonists thirty kilometers south of Colomb-Béchar.

CITRACIT was a hubristic project that never materialized. Citroën abandoned the initiative and ordered his personnel to repatriate, along with all transportable equipment and materials. His decision to cancel CITRACIT lost him over 15 million francs.[104] Forced to give up on CITRACIT, Citroën turned to making

99. CATB, *La Compagnie Transafricaine Citroën (brochure illustré)*; *Le matin*, "Compagnie Transafricaine Citroën."

100. FGMH/MQB, "Lettre de la direction générale usines Citroën au ministre des Colonies," Oct. 6, 1924, DA001110/64069. Citroën "was engaged in direct diplomacy" with King Albert I of Belgium, whom he had met at a screening of the first Saharan crossing film (Denning, "Mobilizing Empire," 62).

101. CATB, *La Compagnie Transafricaine Citroën*, n.d., DOS2016ECR-01057, 133.

102. Murray, "Le tourisme Citroën," 102–5.

103. *Le matin*, "L'inauguration du service automobile Colomb-Béchar est différée."

104. *Le matin*, "Le projet du transafricain automobile est abandonné"; *L'écho de Paris*, "Plus de transafricain automobile!"

his upcoming Sahara expedition a major public relations event. As Murray has pointed out, by manipulating information as well as the film industry and the press, Citroën diverted the attention of metropolitan audiences from the failure of CITRACIT.[105] The trans-Saharan expedition, which had been initially devised as a surveying mission to prepare the groundwork for his grand scheme of a trans-African communications network, was dissociated from CITRACIT and advertised as a stand-alone project.

Galvanized by the success of his first desert crossing, Citroën turned to bigger and even more ambitious plans. He envisaged a grand commercial project connecting traffic across France's colonial possessions, as well as another monumental tourism enterprise offering upmarket travelers unique opportunities to discover French Africa. The failure of his larger initiative did not dissuade him from going forward with a second trans-Saharan expedition, which crossed the continent from Colomb-Béchar to the Indian Ocean and the island of Madagascar. Whether successes or failures, Citroën's hubristic colonial schemes underscore his instrumental, yet unofficial, role in building automobility in French Africa.

Conclusion

As soon as World War I ended, a host of actors—business leaders, entrepreneurs, and industrialists—began to promote driving for pleasure in the Maghreb, creating a colonial model of auto-tourism. Technology's intimate relationship with empire meant that increasingly efficient transportation expanded the geographic range of travel destinations, as did growing French control and influence overseas. This article has argued that, while empire enabled the spread of interwar tourism, automobile tourism in turn constructed the empire.

The article makes this argument through a close focus on the individual companies, entrepreneurs, and industrialists who fomented automobile tourism in North Africa. These include the tire companies Dunlop and Michelin, whose guidebooks equipped tourists with a new genre of travel writing that effectively advocated the empire. Dal Piaz stands out as a central actor among the entrepreneurs who promoted leisured automobility in the Maghreb. Along with the travel guides mentioned above, his CGT created critical infrastructure for mobility in French North Africa. Finally, Citroën's aptitude for recognizing potentially profitable commercial projects made him an influential promoter of travel and tourism in colonial Africa. Along with others, he helped transform the French imperial imagery to include visions of a motorized and mobile continent.

---

105. Murray, "Le tourisme Citroën," 96, 105–6.

The focus on entrepreneurs and companies is not incidental: abandoning monolithic narratives of imperial power, my account highlights the close interconnections among the state, the private sector, and technological development. The entrepreneurs of automobile tourism capitalized on the existence of empire and were directly involved in increasing levels of mobility and accessibility in France's three colonial possessions in North Africa. Their commercial motivations and commodification of the colonies for tourist consumption in turn garnered official support, as the French had recourse to transportation infrastructure and motorization projects for the projection of power within the empire. Taking seriously the material clout and ideological impact of colonial auto-tourism, this article highlights the centrality of the private sector to the functioning of empire while also unpacking the role of automobility in maintaining political structures and economic claims, notably France's interwar imperial objectives.

Finally, this article also tracks a shift in the experience of empire: auto-tourism in interwar North Africa became the juncture point between imperial ideologies and policies. By providing new ways of moving through space, the car enabled tourists to experience the colonies as a spectacle from a series of unique and all-encompassing viewpoints. The automobile thus altered the process of appropriating space, in the same way that Wolfgang Schivelbusch had shown for railways half a century earlier.[106] Motorized traffic became a central component of the politico-administrative apparatus of the interwar French imperial project in North Africa. Colonial auto-tourism encouraged a popular imperial sentiment toward the colonies through consumption and mobility. At its most basic, driving became a mode of symbolic reconquest of colonial territory, a private venture that in turn advanced French influence and interests.

ZOHAR SAPIR DVIR is a PhD candidate at the School of Historical Studies, Tel Aviv University. She is completing her dissertation on the history of automobility in the French imperial Mediterranean during the interwar period.

References

Arnold, David. "Europe, Technology, and Colonialism in the Twentieth Century." *History and Technology* 21, no. 1 (2005): 85–106.

Audouin-Dubreuil, Ariane. *La croisière noire: Sur les traces de l'Expédition Citroën Centre-Afrique.* Glénat, 2014.

Audouin-Dubreuil, Ariane, Etienne Christian, and Marie Christian. *Les croisières Citroën, 1922–1934.* Glénat, 2009.

---

106. Schivelbusch, *Railway Journey*.

Barak, On. "Three Watersheds in the History of Energy." *Comparative Studies of South Asia, Africa and the Middle East* 34, no. 3 (2014): 440–53.

Baranowski, Shelley, Waleed Hazbun, Eric G. E. Zuelow, Christopher Endy, Stephanie Malia Hom, Gordon Pirie, and Trevor Simmons. "Tourism and Empire." *Journal of Tourism History* 7, nos. 1–2 (2015): 100–130.

Bardou, Jean Pierre. *The Automobile Revolution: The Impact of an Industry*. Translated by James Laux. University of North Carolina Press, 1982.

Bejui, Dominique, and Pascal Bejui. *Exploits, fantasmes transsahariens: Quatre-vingts ans de traversées sahariennes abouties ou . . . rêvées, en auto, en camion, en train et en avion*. La Régordane, 1994.

Berthonnet, Arnaud. "Le rôle des militaires français dans la mise en valeur d'un tourisme au Sahara de la fin du XIXe siècle aux années 1930." In *Le tourisme dans l'Empire français: Politiques, pratiques et imaginaire (XIXe—XXe siècles)*, edited by Colette Zytnicki and Habib Kazdaghli, 79–96. La Société française d'histoire d'Outre-mer, 2009.

Bloom, Peter J. *French Colonial Documentary: Mythologies of Humanitarianism*. University of Minnesota Press, 2008.

Bohling, Joseph. "Colonial or Continental Power? The Debate over Economic Expansion in Interwar France, 1925–1932." *Contemporary European History* 26, no. 2 (2017): 217–41.

Brown, Megan. "Le Rallye Méditerranée–Le Cap." *French Politics, Culture, and Society* 38, no. 2 (2020): 80–104.

Chatty, Dawn. *Nomadic Societies in the Middle East and North Africa: Entering the Twenty-First Century*. Brill, 2006.

Clarsen, Georgine, and Lorenzo Veracini. "Settler Colonial Automobilities: A Distinct Constellation of Automobile Cultures?" *History Compass* 10, no. 12 (2012): 889–900.

Cresswell, Tim. "Towards a Politics of Mobility." *Environment and Planning D: Society and Space* 28, no. 1 (2010): 17–31.

Cyril Isnart, Charlotte Mus-Jelidi, and Colette Zytnicki, eds. *Fabrique du tourisme et expériences patrimoniales au Maghreb, XIXe-XXIe siècles*. Centre Jacques-Berque, 2018.

Davis, Diana K. "Restoring Roman Nature: French Identity and North African Environmental History." In *Environmental Imaginaries of the Middle East and North Africa*, edited by Edmund Burke and Diana K. Davis, 60–86. Ohio University Press, 2013.

Demay, Alice, and Caroline Herbelin. "La sala et le micro-palace: L'hôtellerie en Indochine." In *Fabrique du tourisme et expériences patrimoniales au Maghreb, XIXe–XXIe siècles*, edited by Cyril Isnart, Charlotte Mus-Jelidi, and Colette Zytnicki. Centre Jacques-Berque, 2018. https://doi.org/10.4000/books.cjb.1529.

Denning, Andrew. *Automotive Empire: How Cars and Roads Fueled European Colonialism in Africa*. Cornell University Press, 2024.

Denning, Andrew. "Mobilizing Empire: The Citroën Central Africa Expedition and the Interwar Civilizing Mission." *Technology and Culture* 61, no. 1 (2020): 42–70.

Dulucq, Sophie. "'Découvrir l'âme africaine': Les temps obscurs du tourisme culturel en afrique coloniale française (années 1920–années 1950)." *Cahiers d'études africaines*, nos. 193–94 (2009): 27–48.

Edensor, Tim. "Automobility and National Identity: Representation, Geography, and Driving Practice." *Theory, Culture, and Society* 21, nos. 4–5 (2004): 101–20.

Featherstone, Mike, Nigel Thrift, and John Urry, eds. *Automobilities*. Sage, 2005.

Fletcher, Robert. "Running the Corridor: Nomadic Societies and Imperial Rule in the Inter-War Syrian Desert." *Past and Present*, no. 220 (2013): 185–215.

Flonneau, Mathieu. *Les cultures du volant: Essai sur les mondes de l'automobilisme, XXe–XXIe siècles.* Autrement, 2008.

Flonneau, Mathieu, and Vincent Guigueno, eds. *De l'histoire des transports à l'histoire de la mobilité? Etat des lieux, enjeux et perspectives de recherche.* Presses universitaires de Rennes, 2009.

France-Maroc. "La route touristique Alger-Casablanca-Marrakech." Nov. 1920.

Furlough, Ellen. "Making Mass Vacations: Tourism and Consumer Culture in France, 1930s to 1970s." *Society for Comparative Study of Society and History* 40, no. 2 (1998): 247–83.

Furlough, Ellen. "*Une Leçon des Choses*: Tourism, Empire, and the Nation in Interwar France." *French Historical Studies* 25, no. 3 (2002): 441–73.

Geppert, Alexander C. T. *Fleeting Cities: Imperial Expositions in Fin-de-Siècle Europe.* Palgrave Macmillan, 2013.

Grieco, Margaret, and John Urry, eds. *Mobilities: New Perspectives on Transport and Society.* Routledge, 2016.

Gross, Stephen G., and Andrew Needham, eds. *New Energies: A History of Energy Transitions in Europe and North America.* University of Pittsburgh Press, 2023.

Guide Michelin: Maroc, Algérie, Tunisie. 1st ed. Paris, 1927.

Guide Michelin: Maroc, Algérie, Tunisie. 2nd ed. Paris, 1930.

Hannam, Kevin, Mimi Sheller, and John Urry. "Editorial: Mobilities, Immobilities, and Moorings." *Mobilities* 1, no. 1 (2006): 1–22.

Harp, Stephen L. *Marketing Michelin: Advertising and Cultural Identity in Twentieth-Century France.* Johns Hopkins University Press, 2001.

Headrick, Daniel R. *The Tools of Empire: Technology and European Imperialism in the Nineteenth Century.* Oxford University Press, 1981.

Heffernan, Mike. "Shifting Sands: The Trans-Saharan Railway." In *Engineering Earth: The Impacts of Megaengineering Projects*, edited by Stanley D. Brunn, 617–26. Springer, 2011.

Hunter, Robert. "Promoting Empire: The Hachette Tourist in French Morocco, 1919–36." *Middle Eastern Studies* 43, no. 4 (2007): 579–91.

Hunter, Robert. "Tourism and Empire: The Thomas Cook and Son Enterprise on the Nile, 1868–1914." *Middle Eastern Studies* 40, no. 5 (2004): 28–54.

Jackson, Simon. "Personal Connections and Regional Networks: Cross-Border Ford Automobile Distribution in French Mandate Syria." In *Regimes of Mobility: Borders and State Formation in the Middle East, 1918–1946*, edited by Jordi Tejel and Ramazan Hakkı Öztan, 109–40. Edinburgh University Press, 2022.

Koshar, Rudy. "'What Ought to Be Seen': Tourists' Guidebooks and National Identities in Modern Germany and Europe." *Journal of Contemporary History* 33, no. 3 (1998): 323–40.

Larkin, Brian. "The Politics and Poetics of Infrastructure." *Annual Review of Anthropology* 42 (2013): 327–43.

Latour, Bruno. *Reassembling the Social: An Introduction to Actor-Network-Theory.* Oxford, 2005.

Lavenir, Catherine Bertho. "How the Motor Car Conquered the Road." In vol. 9 of *Cultures of Control*, edited by Miriam R. Levin, 113–34. Amsterdam, 2000.

L'écho d'Alger. "Le circuit Nord-Africain: Les touristes sont reçus par le Général Lyautey à Rabat." Oct. 25, 1920.

L'écho d'Alger. "Tourisme." Oct. 10, 1920.

L'écho de Paris. "Plus de transafricain automobile! M. Citroën renonce définitivement à ses grands projets." Jan. 7, 1925.

Le matin. "Compagnie Transafricaine Citroën." Dec. 24, 1924.

Le matin. "Le projet du transafricain automobile est abandonné." Jan. 8, 1925.

Le matin. "L'inauguration du service automobile Colomb-Béchar est différée." Jan. 3, 1925.

Le tourisme automobile au Maroc: Guide Dunlop. Société anonyme des pneumatiques Dunlop, 1922.

Le tourisme automobile en Algérie-Tunisie: Guide Dunlop. Editions des guides du tourisme automobile, 1923.

Loubet, Jean-Louis. *Histoire de l'automobile française.* Editions Seuil, 2001.

Marseille, Jacques. *Empire colonial et capitalisme français: Histoire d'un divorce.* Albin Michel, 1984.

Michelin, André. *Des faits et des chiffres sur l'industrie automobile française.* Michelin, 1929.

Mom, Gijs. *Atlantic Automobilism: Emergence and Persistence of the Car, 1895–1940.* Berghahn Books, 2015.

Mom, Gijs, Peter Norton, Liz Millward, and Mathieu Flonneau, eds. *Mobility in History: Reviews and Reflections.* Presses universitaires suisses, 2011.

Mom, Gijs, Peter Norton, Gordon Pirie, and Georgine Clarsen, eds. *Mobility in History: Themes in Transport.* Presses universitaires suisses, 2010.

Murray, Alison. "Film and Colonial Memory: La Croisière Noire, 1924–2004." In *Memory, Empire, and Postcolonialism: Legacies of French Colonialism,* edited by Alec G. Hargreaves, 81–98. Lexington Books, 2005.

Murray, Alison. "Les automobiles, le désert africain, et le cinéma: Un tourisme imaginaire." In *Le tourisme dans l'Empire français: Politiques, pratiques et imaginaire (XIXe–XXe siècles),* edited by Colette Zytnicki and Habib Kazdaghli, 181–94. La Société française d'histoire d'Outre-mer, 2009.

Murray, Alison. "Le tourisme Citroën au Sahara (1924–1925)." *Vingtième siècle: Revue d'histoire,* no. 68 (2000): 95–107.

Nathan, Roger. "L'industrie automobile." *Revue d'économie politique* 44, no. 3 (1930): 794–823.

O'Connell, Sean. *The Car in British Society: Class, Gender, and Motoring, 1896–1939.* Manchester University Press, 1998.

Olson, Kory. "Come Drive French North Africa: Cartographic and Guidebook Discourse in Michelin's 1929 Maroc, Algérie, Tunisie." *French Colonial History* 20, no. 1 (2021): 29–64.

Peltre, Christine. *Le voyage en Afrique du Nord: Images et mirages d'un tourisme.* Bleu autour, 2018.

Perkins, Kenneth J. "The Compagnie Générale Transatlantique and the Development of Saharan Tourism in North Africa." In *The Business of Tourism: Place, Faith, and History,* edited by Philip Scranton and Janet F. Davidson, 34–55. University of Pennsylvania Press, 2007.

Pétriat, Philippe. "The Uneven Age of Speed: Caravans, Technology, and Mobility in the Late Ottoman and Post-Ottoman Middle East." *International Journal of Middle East Studies* 53, no. 2 (2021): 273–90.

Pinhas, Shira. "The Moment of Oil: Technology and Mobility in Mandate Palestine." PhD diss., Tel Aviv University, 2022.

Pirie, Gordon. "Non-Urban Motoring in Colonial Africa in the 1920s and 1930s." *South African Historical Journal* 63, no. 1 (2011): 38–60.

Rauch, André. *Vacances en France de 1830 à nos jours.* Presses universitaires de France, 1996.

Ricard, Joseph Honoré. *Le grand tourisme dans le nord-africain: L'œuvre du président Dal Piaz.* Ligue maritime et coloniale, 1928.

Sabatès, Fabien. *La croisière noire Citroën, 1924–1925.* Eric Baschet, 1980.

Sachs, Wolfgang. *For Love of the Automobile: Looking Back into the History of Our Desires.* Translated by Don Reneau. University of California Press, 1992.

Schivelbusch, Wolfgang. *The Railway Journey: The Industrialization of Time and Space in the Nineteenth Century.* University of California Press, 1986.

Seiler, Cotten. *Republic of Drivers: A Cultural History of Automobility in America*. University of Chicago Press, 2008.

Sheller, Mimi, and John Urry. "The New Mobilities Paradigm." *Environment and Planning A: Economy and Space* 38, no. 2 (2006): 207–26.

Urry, John. *Mobilities*. Polity, 2007.

Urry, John. "The 'System' of Automobility." *Theory, Culture and Society* 21, nos. 4–5 (2004): 25–39.

Vox, Maximilien. *Les auto-circuits nord-africains*. Le Service typographique,, 1929.

Wilder, Gary. "Framing Greater France Between the Wars." *Journal of Historical Sociology* 14, no. 2 (2001): 198–225.

Young, Patrick. "The Consumer as National Subject: Bourgeois Tourism in the French Third Republic, 1880–1914." PhD diss., University of Michigan, 2001.

Zytnicki, Colette. "'Faire l'Algérie agréable': Tourisme et colonisation en Algérie des années 1870 à 1962." *Le mouvement social*, no. 242 (2013): 97–114.

Zytnicki, Colette, and Habib Kazdaghli, eds. *Le tourisme dans l'Empire français: Politiques, pratiques et imaginaire (XIXe - XXe siècles)*. La Société française d'histoire d'outre-mer, 2009.

Recent Books and Dissertations on French History

Compiled by SARAH SUSSMAN

General and Miscellaneous

Andress, David, ed. *The Routledge Handbook of French History*. London: Routledge, 2024. 647p. $270.00.

Antonutti, Isabelle. *Bâtisseuses de la lecture publique: Une histoire des premières bibliothécaires, 1900–1950*. Papiers. Villeurbanne: Presses de l'Enssib, 2024. 173p. €25.00.

Asgarov, Vazeh. *Immigration des Azerbaïdjanais en France*. Strasbourg: Kapaz, 2022. 472p. €40.00.

Ayling, Lindsay. "Fractured Nationalism and the Crises of French Identity, 1789–1899." Ph.D., The University of North Carolina at Chapel Hill, 2023.

Aynié, Marie, Laurence Croq, and Nicolas Lyon-Caen, eds. *Les religions des Parisiens: D'hier à aujourd'hui*. Paris: Éditions de la Sorbonne, 2024. 499p. €30.00.

Barlow, Michel. *Rue Lanterne: Deux siècles de protestantisme au coeur de Lyon (1832–2022)*. Lyon: Olivétan, 2023. 239p. €27.00.

Belmas, Élisabeth, and Turcot, Laurent, eds. *Jeux, sports et loisirs en France du XVIe au XXe siècle*. Rennes: Presses universitaires de Rennes, 2024. 387p. €28.00.

Bouyssy, Maïté. *Rue Transnonain, 14 avril 1834: Un massacre à la française*. Limoges: Lambert-Lucas, 2024. 288p. €33.00.

Briegel, Françoise, Maria Pia Donato, and Valérie Theis, eds. *Logiques de l'inventaire: Moyen Âge-XIXe siècle*. Rennes: Presses universitaires de Rennes, 2024. 364p. €28.00.

Brown, Ryan, Maximilien Novak, and Colin Jones, eds. *Le pouvoir en procès: Opinion publique et légitimité politique des Lumières au Premier Empire*. Paris: Classiques Garnier, 2024. 211p. €26.00.

Cabanel, Patrick. *La fabrique d'un haut lieu: Le Chambon-sur-Lignon et le plateau: XIXe–XXIe siècle*. Le Cheylard: Dolmazon, 2024. 246p. €30.00.

Chaubet, François, Charlotte Faucher, Laurent Martin, and Nicolas Peyre, eds. *Histoire(s) de la diplomatie culturelle français: Du rayonnement à l'influence*. Toulouse: Editions de l'attribut, 2024. 622p. €19.00.

Demartini, Anne-Emmanuelle, ed. *Alain Corbin: Écrivain de l'histoire*. Paris: Éditions Flammarion, 2024. 297p. €12.00.

Dessaux, Pierre-Antoine. *Vermicelles et coquillettes: Histoire d'une industrie alimentaire*. Tours, France: Presses universitaires François-Rabelais de Tours, 2023. 308p. €28.00.

French Historical Studies • Vol. 48, No. 2 (May 2025) • DOI 10.1215/00161071-11626729

Dufour, Jean-Yves, ed. *Archéologie et histoire des jeux de paume en France: De Versailles à la Marseillaise, XVIe–XVIIIe siècle*. Paris: CNRS Editions: INRAP, 2024. 300p. €49.00.

Dunlop, Catherine Tatiana. *The Mistral: A Windswept History of Modern France*. Chicago: The University of Chicago Press, 2024. 192p. $32.50.

Franz, Laurent. *Charlemagne-Émile de Maupas (1818–1888): Étude d'une trajectoire administrative, politique et notabiliaire, des monarchies censitaires à la troisième République*. Thèmes et commentaires. Bibliothèque parlementaire et constitutionnelle. Paris: Dalloz, 2024. 712p. €79.00.

Gaillard, Claire-Lise. *Pas sérieux s'abstenir: Histoire du marché de la rencontre, XIXe–XXe siècle*. Paris: CNRS, 2024. 350p. €25.00.

Genin, Vincent. *Histoire intellectuelle de la laïcité: De 1905 à nos jours*. Paris: Presses universitaires de France, 2024. 344p. €25.00.

Guicheteau, Samuel, Manuella Noyer, and Christophe Patillon. *Dockers, une histoire nantaise: Travailler et lutter sur les quais (XVIe–XXe siècle)*. Nantes: Éditions CHT, 2023. 293p. €22.00.

Jacquot, Lionel, Brice Monier, Martine Paindorge, and Simon Paye, eds. *Bataville (1931–2001): ville-usine de la chaussure*. Fontaine: PUG, 2023. 379p. €35.00.

Joly, Hervé. *Histoire de l'Ecole polytechnique*. Paris: La Découverte, 2024. 128p. €11.00.

Kaplan, Steven Laurence. *Transmettre, soumettre, socialiser: Essai sur l'apprentissage de Colbert à la Grande Guerre*. Paris: Fayard, 2023. 899p. €38.00.

Lafrance, Xavier, and Stephen Miller. *The Transition to Capitalism in Modern France: Primitive Accumulation and Markets from the Old Regime to the Post-WWII Era*. Abingdon, Oxon: Routledge, 2024. 2236p. $144.00.

Lee, Min Kyung. *The Tyranny of the Straight Line: Mapping Modern Paris*. New Haven: Yale University Press, 2024. 208p. $65.00.

Le Tourneau, Dominique, ed. *Panégyriques de Jeanne d'Arc*. Paris: Honoré Champion, 2024. 3 vols. €265.00.

Maire, Catherine-Laurence, Bernard Bourdin, and Patrice Gueniffey, eds. *Le "gallicanisme": Une singularité française?* Paris: Cerf, 2023. 233p. €22.00.

Malabou, Catherine. *Il n'y a pas eu de Révolution: Réflexions sur la propriété, le pouvoir et la condition servile en France*. Paris: Payot & Rivages, 2024. 316p. €20.00.

Marcelino, Taryn. "Mémoire et Patrimoine: The Present-Day Impact of the History of Slavery in France." Ph.D., University of California, Los Angeles, 2024.

Martinet, Aline. *Enfermer et punir: Histoire des prisons et des prisonniers des Alpes-Maritimes (1792–1939)*. Paris: Classiques Garnier, 2024. 1108p. €45.00.

Maury, Emmanuel. *Le roman secret de Chantilly: Mille ans d'histoire de France*. Paris: Perrin, 2024. 413p. €25.00.

Mermaz, Louis, and Napoli, Aude, eds. *Le fonds Louis Mermaz: Une source pour l'histoire d'une période de mutation de la vie politique française*. Lormont: le Bord de l'eau, 2024. 192p. €22.00

Nenon, Jean-Pierre. *L'ardoise et les ardoisiers de France. Un patrimoine millénaire menacé*. Rennes: Presses universitaires de Rennes, 2024. 248p. €34.00.

Normand, Eric, and Alain Champagne, eds. *De l'eau, du sel et des hommes: Les marais charentais au Moyen Âge et à l'époque moderne, histoire, archéologie, environnement*. Rennes: Presses universitaires de Rennes, 2024. 254p. €35.00.

Porte, Rémy. *Dictionnaire d'histoire militaire de la France: Des origines à nos jours.* Chamalières: LEMME edit, 2024. 600p. €23.00.

Powers, Rebecca Terese. *Balzac on the Barricades: The Literary Origins of an Economic Revolution.* Charlottesville: University of Virginia Press, 2024. 228p. $115.00.

Pozzi, Jérôme. *Approches politiques et culturelles de la France et de la Lorraine. Mélanges en l'honneur de Jean El Gammal.* Nancy: Université de Lorraine, 2024. 352p. €24.00.

Priotti, Jean-Philippe, and Jean-Louis Podvin, eds. *Un siècle d'or culturel en province: Boulogne-sur-Mer entre 1820 et 1920.* 2024. Villeneuve-d'Ascq: Presses universitaires du Septentrion., 2024. 542p. €35.00.

Raffarin, Jean-Pierre. *Etre le numéro deux: Une histoire des rapports de pouvoir à la tête de l'Etat, suivi d'un entretien avec Jean-Pierre Raffarin.* Rennes: Presses universitaires de Rennes, 2024. 312p. €28.00.

Rolloy, Gérard. *Édouard Quesnel (1781–1850): Négociant, armateur et planteur: Le Havre, New York, Cayenne.* Paris: Les Indes savantes, 2024. 475p. €35.00.

Touzery, Mireille. *Payer pour le roi: La fiscalité monarchique (France, 1302–1792).* Ceyzérieu: Champ Vallon, 2024. 1371p. €48.00.

Tribillon, Justinien. *The Zone: An Alternative History of Paris.* London: Verso, 2024. 204p. $29.95.

Zay, Jean. *Jeunesse de la République.* Paris: Bouquins, 2024. 1154p. €33.00.

Zdatny, Steven. *A History of Hygiene in Modern France: The Threshold of Disgust.* London: Bloomsbury Academic, 2024. 315p. $115.00.

Medieval and Renaissance

Alonge, H. O., Nicolas Balzamo, and Jean Sénié, eds. *Oltralpe: Acteurs, idées et livres entre France et Italie au XVIe siècle.* Roma: Viella, 2023. 358p. €38.00.

Arnold, John. *The Making of Lay Religion in Southern France, c. 1000–1350.* Oxford: Oxford University Press, 2024. 524p. $170.00.

Bailly-Maître, Marie-Christine, ed. *Vivre en montagne au Moyen Âge: Les objets racontent l'histoire de l'argenteria de Brandis, Huez-Alpe d'Huez, XIIe-XIVe siècles.* Lyon: CIHAM-Édition, 2024. 501p. €46.00.

Baudin, Arnaud, Valérie Toureille, and Jean-Marie Yante. *Guerre et paix en Champagne à la fin du Moyen-Âge. Autour du traité de Troyes.* Gent: Snoeck Ducaju & Zoon, 2023. 483p. €30.00.

Bresc, Henri. *Au pays des villages perchés: Une histoire des villages de Provence au Moyen Âge, du pays de Fayence à La Napoule.* Paris: Scudéry, 2024. 623p. €29.90.

Carbonne, Raphael. *Les messagers à la cour de Philippe le Bon et Charles le Téméraire.* Paris: Harmattan, 2024. 191p. €21.00.

Dejoux, Marie, Pierre-Anne Forcadet, Vincent Martin, and Liêm Tuttle. *La justice de Saint Louis: Dans l'ombre du chêne.* Paris: Presses universitaires de France, 2024. 397p. €28.00.

Donnell, Natalie. "Noblewomen's Political Networks Across the European Wars of Religion (1559–1633)." Ph.D., Georgetown University, 2024.

Firnhaber-Baker, Justine. *House of Lilies: The Dynasty That Made Medieval France.* London: Allen Lane, 2024. 408p. $35.00.

Goldman, Oury. *L'empreinte des lointains: Traduire les savoirs sur le monde en France au XVIe siècle.* Geneva: Droz, 2024. 593p. €44.00.

Hamilton, Tom. *A Widow's Vengeance after the Wars of Religion: Gender and Justice in Renaissance France*. Oxford: Oxford University Press, 2024. 240p. $100.00.

Laffont, Pierre-Yves, Bachelier, Julien, Chollet, Samuel, and Meuret, Jean-Claude, eds. *Les mondes ruraux de l'ouest de la France au Moyen Age: Société, pouvoirs, habitats: Études offertes à Daniel Pichot*. Rennes: Presses universitaires de Rennes, 2024. 363p. €25.00.

Lett, Didier. *Crime, genre et châtiment*. Malakoff: Armand Colin, 2024. 448p. €37.00.

Mathieu, Isabelle, and Thierry Pécout, eds. *Un Moyen Âge en partage: Hommage à Jean-Michel Matz*. Rennes: Presses universitaires de Rennes, 2024. 480p. €32.00.

Metz à la fin du Moyen Âge: Fin XIVe-milieu XVIe siècle. Cinisello Balsamo (Milan): Silvana, 2024. 432p. €25.00.

Nofri, Gaël. *Bouvines: La confirmation de la souveraineté*. Paris: Passés composés, 2024. 237p. €20.00.

Oddy, Niall. *Writing Europe in Renaissance France: Travels in Reality and Imagination*. Edinburgh: Edinburgh University Press, 2024. 176p. $100.00.

Santamaria, Jean-Baptiste. *Gouverner au féminin: Marguerite de France, Comtesse d'Artois et de Bourgogne, 1361–1382*. Villeneuve d'Ascq, France: Presses universitaires du Septentrion, 2024. 454p. €32.00.

Stephens, Leigh Vella. "Violating the Body Politic: The Politics of Suffering, Gender, and Royal Authority During the French Wars of Religion (1560–1589)." Ph.D., Georgetown University, 2024.

Szabari, Antónia. *Agents without Empire: Mobility and Race-Making in Sixteenth-Century France*. New York: Fordham University Press, 2024. 294p. $125.00.

Ancien Régime

Avezou, Laurent. *Sully: Bâtisseur de la France moderne*. Paris: Tallandier, 2024. 603p. €26.90.

Banks, Bryan. *Write to Return: Huguenot Refugees on the Frontiers of the French Enlightenment*. Montreal: McGill-Queen's University Press, 2024. 216p. CAD$110.00.

Baudoin, Marie. *The Art of Childbirth: A Seventeenth-Century Midwife's Epistolary Treatise to Doctor Vallant*. Edited by Cathy McClive. A bilingual edition. New York: Iter Press, 2022. 254p. $54.95.

Boscani, Simona, Claire Gantet, and André Holenstein. *Le Corps helvétique et la France (1660–1792): Transferts, asymétries et interdépendances entre des partenaires inégaux = Das Corps helvétique und Frankreich: Transfers, Asymmetrien und Interdependenzen zwischen ungleichen Partnern*. Travaux sur la Suisse des Lumières. Genève: Slatkine, 2024. 394p. €50.00.

Buffet, Marguerite. *New Observations on the French Language; with Praises of Illustrious Learned Women*. Edited by Lynn S. Meskill. New York: Iter Press, 2023. 139p. $43.95.

Charles-Dominique, Luc. *Musiques et musiciens des fêtes urbaines et villageoises en France (XIVe - XVIIIe siècle)*. Turnhout: Brepols, 2024. 1063p. €165.00.

Condren, John. *Louis XIV and the Peace of Europe: French Diplomacy in Northern Italy, 1659–1701*. New York: Routledge, 2024. 243p. $180.00.

Dubuisson, Jean. *Histoire de Madame de Rus: De Beaumes-de-Venise à la cour de France*. Les Beaumes-de-Venise: Les Éditions de l'Académie de Beaumes-de-Venise, 2021. 107p. €17.50.

Gantet, Claire. *La guerre de Trente Ans: 1618–1648*. Paris: Tallandier; Ministère des Armées, 2024. 634p. €26.90.

Huchet de Quénétain, Christophe, Michel David-Weill, and Michèle Bimbenet-Privat, eds. *Nicolas Besnier (1686–1754): Architecte, orfèvre du roi et échevin de la Ville de Paris*. Rennes: Presses universitaires de Rennes, 2024. 348p. €39.00.

Kandakou, Dzianis, and Alexandre Stroev. *Russes à Paris au XVIIIe siècle sous l'oeil de la police*. Paris: Harmattan, 2024. 736p. €60.00.

Lestienne, Cécile. *La salle à manger des Lumières: Histoire, architecture, décor*. Tours: Presses universitaires François-Rabelais, 2024. 281p. €26.00.

Ludington, Charles C. *The Irish in Eighteenth-Century Bordeaux: Contexts, Relations, and Commodities*. London: Routledge, 2023. 262p. $180.00.

Merrick, Jeffrey, ed. *Policing Same-Sex Relations in Eighteenth-Century Paris: Archival Voices from 1785*. University Park: Pennsylvania State University Press, 2024. 251p. $124.95.

Miceli, Erick. *Les révolutions corses et l'idée républicaine: Pascal Paoli face à ses innovations, limites et contradictions (1755–1769)*. Bordeaux: Le Bord de l'eau, 2024. 329p. €26.00.

Montenach, Anne. *Gender, Space, and Illicit Economies in Eighteenth-Century Europe: Uncontrolled Crossings*. London: Routledge, 2024. 290p. $170.00.

Murphy, Neil. *Plague, Towns, and Monarchy in Early Modern France*. Elements in the Renaissance. Cambridge: Cambridge University Press, 2024. 84p. $22.00.

Nelson, William Max. *Enlightenment Biopolitics: A History of Race, Eugenics, and the Making of Citizens*. Chicago: The University of Chicago Press, 2024. 311p. $105.00.

Nicol, Charles. *Les Protestants et Nantes: Une minorité au cœur d'une ville portuaire*. Rennes: Presses universitaires de Rennes, 2024. 301p. €23.00.

Orain, Arnaud. *The Politics of Utopia: A New History of John Law's System, 1695–1795*. Translated by Andrew Brown. Chicago: The University of Chicago Press, 2024. 344p. $45.00.

Roberts, Chloé. "The Demonological Republic of Letters: Judges, Lawyers, Elites, and Demonologists in Early Modern Europe." Ph.D., University of California, Santa Barbara, 2024.

Rovere, Ange. *Pascal Paoli: De Lumières et d'ombres*. Paris: Classiques Garnier, 2024. 483p. €48.00.

Scuiller, Sklaerenn. *Les échanges du quotidien: Le commerce alimentaire en Bretagne au XVIIIe siècle*. Rennes: Presses universitaires de Rennes, 2024. 341p. €26.00.

Tatoueix, Laura. *Défaire son fruit: Une histoire sociale de l'avortement en France à l'époque moderne*. Paris: Éditions EHESS, 2024. 391p. €22.80.

French Revolution and Napoleonic Era

Alpaugh, Micah. *The People's Revolution of 1789*. Ithaca: Cornell University Press, 2024. 402p. $60.95.

Alzas, Nathalie. *Marianne aux enfers: La haine de la Révolution française*. Paris: Editions Critiques, 2024. 208p. €18.00.

Barbaray, Côme. *La guerre de siège à l'épreuve de la Révolution française: Retentissements et perceptions*. Mont-Saint-Aignan: Presses universitaires de Rouen et du Havre, 2024. 200p. €17.00.

Blackmore, Callum. "Opera at the Dawn of Capitalism: Staging Economic Change in France and Its Colonies From the Regency to the Terror." Ph.D., Columbia University, 2024.

Callaway, H. B. *The House in the Rue Saint-Fiacre: A Social History of Property in Revolutionary Paris*. Cambridge, MA: Harvard University Press, 2023. 297p. $45.00.

Croft, Marissa G. "State of Dress: Rhetoric and Clothing Reform in Revolutionary France, 1789–1804." Ph.D., Northwestern University, 2024.

Duplay, Élisabeth. *Femme de révolutionnaire: D'après les Mémoires d'Élisabeth Duplay (1773–1859), veuve Le Bas.* Chamalières: LEMME edit, 2024. 165p. €19.00.

Duroc, Géraud Christophe Michel. *Correspondance du grand maréchal du palais de Napoléon Ier.* Edited by Jean-Pierre Samoyault and Charles-Éloi Vial. Paris: Honoré Champion, 2023. 1301p. €98.00.

Empereur-Mot, Théo. *L'émigration en Savoie sous la Révolution française: Exils et réintégrations, 1792–1818.* Chambéry: Société savoisienne d'histoire et d'archéologie, 2024. 185p. €25.00.

Grall, Jean-Jacques. *Le blocus: Napoléon et le blocus maritime: Pointe de Bretagne 1793–1815.* Châteaulin: Locus solus, 2024. 254p. €24.00.

Hardman, John. *Barnave: The Revolutionary Who Lost His Head for Marie-Antoinette.* New Haven: Yale University Press, 2023. 416p. $40.00.

Higonnet, Anne. *Liberty Equality Fashion: The Women Who Styled the French Revolution.* New York: W.W. Norton & Company, 2024. 286p. $35.00.

Houdecek, François, Michel Roucaud, and Nicolas Texier, eds. *Les grognards ont la parole: Correspondances inédites de soldats et officiers du premier Empire conservées au Service historique de la Défense (1794–1815).* Paris, Vincennes: Editions Pierre de Taillac; Service historique de la Défense, 2024. 400p. €26.90.

Le Carvèse, Patrick. *Les premiers préfets maritimes, 1800–1815.* Paris: SPM, 2024. 1055p. €55.00.

Lheureux-Prévot, Chantal. *Le sexe contrôlé: Être femme après la Révolution (1800–1815).* Paris: Passés composés, 2024. 374p. €24.00.

Malcolm, Hannah N. "Archiving the French Revolution." Ph.D., Indiana University, 2024.

Marini, Giuseppe. *Napoleone e la guerra del 1809 in Italia.* Zeta rifili. Pasian di Prato (UD) Italia: Campanotto editore, 2024. 173p. €20.00.

Martin, Jean-Clément. *La Grande Peur de juillet 1789.* Paris: Tallandier, 2024. 409p. €22.90.

Michelet, Maxime, ed. *Eugénie, impératrice des Français: Actes du colloque du centenaire de la disparition d'Eugénie de Montijo.* Paris: Cerf, 2024. 232p. €29.00.

Novak, Maximilien, and Ryan Brown. *Le pouvoir en procès. Opinion publique et légitimité politique des Lumières au Premier Empire.* Paris: Classiques Garnier, 2024. 211p. €28.60.

Petit, Vincent. *L'adieu au serment: Prêtres et religieux rétractés du diocèse de Besançon (1791–1800).* Besançon: Cêtre, 2024. 305p. €28.00.

Petitier, Paule, ed. *Déchiffrer la tempête: Michelet et la Révolution française.* Rennes: Presses universitaires de Rennes, 2024. 343p. €25.00.

Pilbeam, Pamela M. *The Revolting French, 1787–1889.* London: Routledge, 2024. 160p. $180.00.

Rousseliere, Genevieve. *Sharing Freedom: Republicanism and Exclusion in Revolutionary France.* New York, NY: Cambridge University Press, 2024. 256p. $105.00, $34.99pb.

Sarkozy, Louis. *Napoleon's Library: The Emperor, His Books and Their Influence on the Napoleonic Era.* Barnesly, UK: Frontline Books, 2024. 263p. $42.95.

Thompson, Victoria, Bryant T. Ragan, and Suzanne Desan, eds. *Everyday Politics and Culture in Revolutionary France: Essays in Honor of Lynn Hunt.* Liverpool: Liverpool University Press on behalf of Voltaire Foundation, 2024. 299p. $99.00.

Volney, C.-F. *Volney: "The Ruins" and "Catechism of Natural Law."* Edited by Colin Kidd. Translated by Lucy Kidd. Cambridge: Cambridge University Press, 2024. 207p. $105.00. $34.99pb.

Wahnich, Sophie. *La Révolution des sentiments: Comment faire une cité, 1789–1794.* Paris: Seuil, 2024. 391p. €24.00.

Alimi-Levy, Yohanna. *La démocratie américaine et les révolutions françaises de 1830 et 1848*. Paris: Sorbonne université presses, 2023. 474p. €22.00.

Anceau, Éric, ed. *Les quarante-huitards et les autres: Dictionnaire des dirigeants de 1848*. Paris: Sorbonne Université Presses, 2024. 1778p. €60.00.

Beizer, Janet L. *The Harlequin Eaters: From Food Scraps to Modernism in Nineteenth-Century France*. Minneapolis: University of Minnesota Press, 2024. 331p. $120.00.

Bonin, Hubert. *De la Banque de Bordeaux à la Banque de France (1818–1848–1854): Un demi-siècle de banque d'émission et d'escompte*. Bordeaux: Memoring, 2023. 229p. €22.00.

Bouchet, Thomas. *L'aiguille et la plume: Jules Gay, Désirée Véret, 1807–1897*. Paris: Anamosa, 2024. 622p. €28.00.

Brandely, Emmanuel. *Des historiens contre la Commune*. Paris: Les Nuits rouges, 2024. 216p. €15.00.

Calamita, Umberto. *La Comune di Parigi: In dodici conversazioni radiofoniche e sei appendici*. Vetralla: Davide Ghaleb editore, 2022. 206p. €20.00.

Chavannes, Herminie. *C'est bien dans la Babylone moderne que je me rends seule: Journal d'un voyage à Paris en 1827*. Lausanne: Éditions d'en bas, 2023. 165p. €18.00.

Courtney, Cecil P., Paul Rowe, and Dominique Triaire, eds. *Correspondance générale, 1825–1826*. Berlin: De Gruyter, 2023. 575p. $260.00.

Dandois, Bernard, and Philippe Buonarroti. *Philippe Buonarroti: un révolutionnaire professionnel à Bruxelles (1824–1830): Récit épistolaire*. Bruxelles: Édition Samsa, 2023. 310p. €26.00.

Davies, Helen M. *Herminie and Fanny Pereire: Elite Jewish Women in Nineteenth-Century France*. Manchester: Manchester University Press, 2024. 336p. £85.00.

Dittmar, Gérald. *Albert Robida: Paris pendant la Commune de 1871: Journal et dessins*. Ouistreham: Dittmar, 2024. 208p. €30.00.

Englert, Gianna. *Democracy Tamed: French Liberalism and the Politics of Suffrage*. New York, NY: Oxford University Press, 2024. 213p. $90.00.

Godin, Christian. *Victor Hugo et la Commune*. Ceyzérieu: Champ Vallon, 2024. 403p. €27.00.

Graber, Frédéric. *L'affichage administratif au XIXe siècle: Former le consentement*. Paris: Éditions de la Sorbonne, 2023. 399p. €25.00.

Hayat, Samuel. *Revolutionary Republicanism Participation and Representation in 1848 France*. London: Routledge, 2024. 229p. $180.00.

Levitt, Theresa. *Elixir: A Parisian Perfume House and the Quest for the Secret of Life*. Cambridge, MA: Harvard University Press, 2023. 320p. $32.95.

Lubliner-Mattatia, Sabine. *Le Bronzeland parisien: Un monde disparu*. Paris: Les Indes Savantes, 2022. 581p. €38.00.

Mensch, Matthieu. *Les femmes de Louis XVIII*. Paris: Perrin, 2024. 384p. €23.00.

Tomasello, Federico. *The Making of the Citizen-Worker: Labour and the Borders of Politics in Post-Revolutionary France*. Abingdon, Oxon: Routledge, 2024. 162p. $144.00.

Troyansky, David G. *Entitlement and Complaint: Ending Careers and Reviewing Lives in Post-Revolutionary France*. New York, NY: Oxford University Press, 2023. 236p. $83.00.

Varley, Karine, ed. *The Franco-Prussian War: Turning-Points in European Experiences and Perceptions of Military Conflict*. Abingdon, Oxon: Routledge, 2024. 204p. €180.00.

Third Republic

Asleson, Robyn, Zakiya R. Adair, Samuel N. Dorf, Tirza True Latimer, and T. Denean Sharpley-Whiting. *Brilliant Exiles: American Women in Paris, 1900–1939.* Washington, DC, New Haven: National Portrait Gallery; in association with Yale University Press, 2024. 288p. $60.00.

Baecque, Antoine de. *Sports Belle Époque: Naissance de la passion sportive, 1870–1924.* Paris: Passés composés, 2024. 345p. €22.00.

Barey, Morgane. *Enseigner la guerre: Former les chefs,1918–1945.* Paris: Perrin: Ministère des Armées, 2024. 361p. €25.00.

Bouchet, Julien. *Charles de Freycinet: Bâtisseur de la République.* Neuilly: Atlande, 2024. 288p. €25.00.

Caillaux, Joseph. *Mémoires.* Paris: Perrin, 2024. 413p. €25.00.

Chapelle, Sandra. *Des civils au cœur de la guerre franco-allemande: Écritures de soi et expériences sensibles (1870–1914).* Dijon: Éditions universitaires de Dijon, 2024. 267p. €22.00.

Christen, Carole, ed. *Jules Siegfried (1837–1922): Négociant international, républicain libéral, réformateur social.* Paris: Classiques Garnier, 2024. 512p. €45.00.

Dard, Olivier, and Jean Philippet. *Février 34: L'affrontement.* Paris: Fayard, 2024. 746p. €34.00.

Duruflé, Julie, and Etienne Manchette, eds. *1913: Vers la Grande Guerre, archives de presse.* Paris: Tallandier, 2024. 286p. €19.90.

Fabre, Mélanie. *Hussardes noires: Des enseignantes à l'avant-garde des luttes: de l'affaire Dreyfus à la Grande Guerre.* Marseille: Agone, 2024. 432p. €23.00.

Faucher, Charlotte. *Propaganda, Gender, and Cultural Power: Projections and Perceptions of France in Britain c. 1880–1944.* Oxford: Oxford University Press, 2022. 270p. $85.00.

Georges, Raphaël. *Un nouveau départ: Les vétérans alsaciens-lorrains dans la France d'après-guerre, 1918–1939.* Rennes: Presses universitaires de Rennes, 2024. 374p. €26.00.

Glaumaud-Carbonnier, Marion, and Nicholas White. *Lendemains de défaite: 1870–1871 dans l'imaginaire de la IIIe République.* Lyon: Presses universitaires de Lyon, 2024. 231p. €20.00.

Goirand Hohl, Anne, Philippe Péchoux, and Morgan Poggioli. *"Nous ne plierons pas". Marie Guillot: Une institutrice féministe syndicaliste-révolutionnaire (1880–1934).* Dijon: Editions Universitaires Dijon, 2024. 345p. €25.00.

Grout, Holly. *Playing Cleopatra: Inventing the Female Celebrity in Third Republic France.* Baton Rouge: Louisiana State University Press, 2024. 222p. $50.00.

Huc, Martin. *Marseille interdite: 1878–1943, histoire du quartier réservé.* Paris: la Manufacture de livres, 2024. 399p. €25.00.

Ilacqua, Talitha. *Inventing the Modern Region: Basque Identity and the French Nation-State.* Manchester: Manchester University Press, 2024. 261p. £85.00.

Jouffroy, Denis. *L'histoire climatique de la Corse: De la Belle Époque à 1914, jalons pour une histoire environnementale de la Corse.* Ajaccio: Éditions Albiana, 2024. 352p. €25.00.

Kwok, Alice Coulter Main. "Reimagining the Medieval Norse in Nineteenth-Century France." Ph.D., The University of Wisconsin–Madison, 2024.

Le Sonn, Loïg. *Domestiquer le corps social: Expérimentations sur les femmes, les enfants et les aliénés au temps d'Alfred Binet.* Villeneuve-d'Ascq, France: Presses universitaires du Septentrion, 2023. 362p. €25.00.

Llopart, Michaël. *Aux origines d'AZF: Le problème de l'azote en France, 1919–1940.* Tours, France: Presses universitaires François-Rabelais, 2023. 398p. €28.00.

Myrick, Richard. "Action Libérale Populaire and the Legacy of Catholic Republicans in the French Third Republic." Ph.D., Kansas State University, 2023.

Potier, Frédéric. *Jaurès en duel*. Lormont, Paris: le Bord de l'eau, 2024. 150p. €14.00.

Rapport, Michael. *City of Light, City of Shadows: Paris in the Belle Époque*. New York: Basic Books, 2024. 437p. $35.00.

Samuels, Maurice. *Alfred Dreyfus: The Man at the Center of the Affair*. New Haven: Yale University Press, 2024. 224p. $26.00.

Stamler, Hannah M. "Patrimony in Miniature: French Childhood, Culture, and Media in the Shadow of Depopulation, 1900–1940." Ph.D., Princeton University, 2024.

Steuckardt, Agnès, Corinne Gomila, and Chantal Wionet, eds. *Gens ordinaires dans la Grande Guerre: Correspondances, récits, témoignages*. Charenton-le-Pont: Éditions de la Maison des sciences de l'homme, 2024. 304p. €19.00.

Szabari, Antónia. *Agents without Empire: Mobility and Race-Making in Sixteenth-Century France*. New York: Fordham University Press, 2024. 294p. $125.00.

Thivend, Marianne. *Ces femmes qui comptent: Le genre de l'enseignement commercial en France au xixe siècle*. Lyon: Presses universitaires de Lyon, 2024. 320p. €22.00.

Vout, Caroline, and Christopher Young. *Paris 1924. Sport, Art and the Body*. London: Paul Holberton Publishing, 2024. 208p. $40.00.

Zientek, Adam D. *A Thirst for Wine and War: The Intoxication of French Soldiers on the Western Front*. Montreal: McGill-Queen's University Press, 2024. 272p. $110.00.

1940 to the present

Albertini, Pierre. *Giscard: Le président qui osa*. Paris: L'Archipel, 2024. 312p. €22.00.

Badalassi, Nicolas. *La France, la guerre froide et la Méditerranée: Des accords d'Evian à la Perestroïka*. Rennes: Presses universitaires de Rennes, 2024. 340p. €25.00.

Balent, André, and Richard Vassakos, eds. *Réinventer la gauche en Languedoc-Roussillon, 1945–1968: Guerre froide, anticolonialisme, occitanisme et genèse de la nouvelle gauche*. Perpignan: Publications de l'Olivier, 2022. 234p. €22.00.

Bande, Alexandre, Rudy Reichstadt, and Pierre-Jérôme Biscarat, eds. *Histoire politique de l'antisémitisme en France: De 1967 à nos jours*. Paris: Robert Laffont, 2024. 381p. €22.00.

Becker, Annette. *Des Juifs trahis par leur France: 1939–1944*. Paris: Gallimard, 2024. 292p. €22.00.

Bernard, Nicolas. *Oradour-sur-Glane, 10 juin 1944: Histoire d'un massacre dans l'Europe nazie*. Paris: Tallandier: Ministère des Armées, 2024. 393p. €22.90.

Bishop, Patrick. *Paris 1944: Occupation, Resistance, Liberation*. New York: Pegasus Books, 2024. 374p. $35.00.

Bonnotte, Claire, and Christina Kott. *Châteaux et musées franciliens pendant la Seconde Guerre Mondiale: Une protection stratégique*. Paris: Hermann, 2024. 374p. $35.00.

Botrel, Yannick. *Le Bezen Perrot: Supplétifs des nazis en Bretagne, 1943–1945*. Morlaix: Skol Vreizh, 2024. 280p. €20.00.

Brana, Pierre. *Collaboratrices: 1940–1945: Histoire des femmes qui ont soutenu le régime de Vichy et l'occupant nazi*. Paris: Perrin, 2024. 382p. €24.00.

Branca, Éric. *La République des imposteurs: Chronique indiscrète de la France d'aprés-guerre, 1944–1954*. Paris: Perrin, 2024. 311p. €21.00

Brull-Ulmann, Colette, and Jean-Christophe Portes. *Through the Morgue Door: One Woman's Story of Survival and Saving Children in German-Occupied Paris*. Translated by Anne Landau and Margaret Sinclair. Philadelphia: University of Pennsylvania Press, 2024. 236p. $34.95.

Cordier, Daniel. *Alias Caracalla*. Translated by Rupert Swyer. Chicago: Swan Isle Press, 2024. 776p. $40.00.

Duclert, Vincent, ed. *En attendant la victoire: Messages au général de Gaulle et à la France libre*. Paris: Gallimard, 2024. 272p. €9.40.

Empaytaz, Frédéric. *Frédéric Empaytaz: Dernier préfet du Lot nommé par le gouvernement de Vichy (mars-août 1944), témoignage*. Arcambal: Edicausse, 2024. 235p. €30.00.

Faucher, Charlotte, Laure Humbert, Thomas Piketty, and Thomas Vaisset, eds. *Françaises et Français libres: Une identité née de la pluralité*. Rennes: Presses universitaires de Rennes, 2024. 230p. €22.00.

Fedorka, Drew Michael. "France Remade: Youth, Modernization, and the Politics of Progress During the Trente Glorieuses." Ph.D., New York University, 2024.

Fosseux, Marc. *De Gaulle et Paris*. Paris: Nouveau Monde, 2023. 371p. €25.90.

Gallot, Fanny. *Mobilisées!: Une histoire féministe des contestations populaires*. Paris: Seuil, 2024. 283p. €22.50.

Grillère, Diane. *L'autre Occupation: l'Italie fasciste en France, 1940–1943*. Paris: Nouveau Monde, 2023. 626p. €25.90.

Guedj, Jérémy. *Les Juifs français et le nazisme (1933–1939): L'histoire renversée*. Paris: Presses universitaires de France, 2024. 392p. €22.00.

Harrison, Olivia C. *Natives against Nativism: Antiracism and Indigenous Critique in Postcolonial France*. Minneapolis, MN: University of Minnesota Press, 2023. 284p. $112.00/$28.00pb.

Leray, Gérard. *Les derniers jours de Jean Moulin à Chartres*. Lèves: Ella éditions, 2020. 263p. €19.00.

Lozac'h, Alain. *Résister En Bretagne: Les Combattants volontaires de la Résistance des Côtes-Du-Nord*. Morlaix: Skol Vreizh, 2024. 360p. €20.00.

Lugassy, Maurice. *L'Armée juive, une résistance française: Ces combattants au service de la Libération*. Toulouse: Privat, 2024. 316p. €19.90.

Malassis, Frantz. *Histoire d'objets de la Résistance*. Paris: Histoire & Collections, 2024. 111p. €22.00.

Manigand, Christine, and Olivier Sibre, eds. *Le dictionnaire Pompidou*. Paris: Robert Laffont, 2024. 731p. €29.50.

Marcil-Bergeron, Myriam. *Le chant des sirènes: Récits d'exploration sous-marine en France (1950–1960)*. Montréal: Presses de l'Université de Montréal, 2023. 264p. $34.95CAD

Martignoles, Nicolas. *Foges: Histoire et mémoire d'un combat de la Résistance*. Marmande: Jacques André éditeur, 2023. 128p. €16.00.

Miot, Claire. *Le débarquement de Provence: Août 1944*. Paris: Passés composés: Ministère des Armées: ECPAD, 2024. 175p. €25.00.

Muracciole, Jean-François. *Quand De Gaulle libère Paris: Juin-août 1944: récit d'une prise de pouvoir*. Paris: Odile Jacob, 2024. 491p. €24.90.

Neiberg, Michael S. *When France Fell: The Vichy Crisis and the Fate of the Anglo-American Alliance*. Cambridge, MA: Harvard University Press, 2023. 312p. $21.95.

Niget, David, and Jean-Luc Marais. *Cloîtrées. Filles et religieuses dans les internats de rééducation du Bon-Pasteur d'Angers, 1940–1990*. Rennes: Presses universitaires de Rennes, 2024. 328p. €25.00.

Raymond, Gino. *France since the Liberation: Between Exceptionalism and Convergence*. Abingdon, Oxon: Routledge, 2024. 266p. $144.00.

Rossi, Emanuele R.C. *La vendetta della 4. Repubblica: l'annessione di Tenda e Briga Marittima*. Roma: Nuova Cultura, 2023. 167p. €23.00.

Torres, Félix. *Le décrochage français: Histoire d'une contreperformance politique et économique (1983–2017)*. Paris: Presses universitaires de France, 2024. 522p. €26.00.

Tribillon, Justinien. *The Zone: An Alternative History of Paris*. London: Verso, 2024. 204p. $29.95.

Vergez-Chaignon, Bénédicte. *The Man Who Murdered Admiral Darlan: Vichy, the Allies and the Resistance in French North Africa*. London: Routledge, 2023. 228p. $144.00.

Weckel, Michel. *Entre silences et non-dits, les protestants d'Alsace face au nazisme*. Strasbourg: Nuée bleue, 2024. 162p. €22.00.

France and the World

Alencar Gálvez, Raúl. "Dangerous Liaisons: Merchant Societies and Trading Networks Between Peru and France, 1698–1764." Ph.D., Tulane University, 2024.

Amar, Marianne, Aline Angoustures, Dzovinar Kévonian, and Anouche Kunth, eds. *Entre décolonisation et guerre froide: Administrer l'asile, 1960–1990*. Rennes: Presses universitaires de Rennes, 2024. 398p. €28.00.

Andurain, Julie d'. *Les troupes coloniales: Une histoire politique et militaire*. Paris: Passés composés, 2024. 394p. €23.50.

Barnes, Whitney Abernathy. "Remaking Religion: Islam, Empire, Race, and the Secularization of French Christianity 1830–1920." Ph.D., Boston College, 2023.

Berlemont, Johanne, ed. *Combattre loin de chez soi: L'empire colonial français dans la Grande Guerre: Exposition, Meaux, Musée de La Grande Guerre, 6 avril–30 décembre 2024*. Paris: In fine; Musée de la Grande guerre, pays de Meaux, 2024. 200p. €25.00.

Bertrand, Romain. *Les grandes déconvenues: La Renaissance, Sumatra, les frères Parmentier*. Paris: Seuil, 2024. 380p. €24.50.

Blanchard, Emmanuel. *Des colonisés ingouvernables: Adresses d'Algériens aux autorités françaises (Akbou, Paris, 1919–1940)*. Paris: Presses de Sciences Po, 2024. 260p. €26.00.

Boyle, Catherine Fitzgerald. "Sites of Servitude: Comparative Slavery, Difference, and Belonging in Tunis, 1736–1891." Ph.D., Harvard University, 2024.

Dao, Sirui. "Travels and Investigations in the Yunnan-Burma Borderlands, 1837–1911." Ph.D., Universität Hamburg (Germany), 2023.

Destaney, Emmanuel. *America's French Orphans: Mobilization, Humanitarianism, and the Protection of France, 1914–1921*. Cambridge, United Kingdom; New York, NY: Cambridge University Press, 2024. 281p.£47.99

Glover, Tayzhaun. "Freedom on the Horizon: Transmarine Marronage and the Abolition of Slavery in Dominica, Martinique, and St. Lucia, 1824–1848." Ph.D., Duke University, 2024.

Harrison, Carol E., and Thomas J. Brown. *Zouave Theaters: Transnational Military Fashion and Performance*. Baton Rouge: Louisiana State University Press, 2024. 330p. $50.00.

Harvey, David Allen. *Tropical Despotisms: Enlightened Reform in the French Caribbean*. Ithaca: Cornell University Press, 2024. 306p. $64.95.

Klooster, Wim, ed. *The Cambridge History of the Age of the Atlantic Revolutions. Volume II, France, Europe, and Haiti*. Cambridge, UK: Cambridge University Press, 2023. 790p. $150.00.

Kramer, Lloyd S. *Traveling to Unknown Places: Nineteenth-Century Journeys toward French and American Selfhood*. Chapel Hill: The University of North Carolina Press, 2024. 384p. $99.00.

Lange, Erik de. *Menacing Tides: Security, Piracy and Empire in the Nineteenth-Century Mediterranean*. Cambridge: Cambridge University Press, 2024. 335p. $110.00.

Ledoux, Sébastien, and Paul Max Morin. *L'Algérie de Macron: les impasses d'une politique mémorielle*. Paris: Presses universitaires de France, 2024. 294p. €20.00.

Luneau, Tyson A. "A Reimagined Periphery: Environment and the Construction of the French Colonial Empire in North Africa." Ph.D., State University of New York at Albany, 2024.

Maillard, Bruno. *La vie des esclaves en prison: La Réunion, 1767–1848*. Paris: Plon, 2024. 422p. €23.90.

Marey-Monge, Alphonse, Alphonse. *Journal, Volume 1: Voyages en Chine et aux colonies (1841 à 1845)*. Edited by Myriam Tsimbidy. Paris: Classiques Garnier, 2024. 2 vols, 1578p. €98.00.

McLeod, Jane. *Print, Politics and Trade in the French Atlantic: The Labottière Family as Eighteenth-Century Cultural Brokers*. Woodbridge, Suffolk: The Boydell Press, 2024. 348p. £95.00.

Meslien, Sylvie, Éliane Sempaire-Étienne, and Sylvain Demange. *1940–1943, résistances & dissidences aux Antilles et en Guyane: Mémoires de guerre, mémoires de vie*. Saint-Denis (La Réunion): Orphie, 2023. 288p. €26.00.

Mormin-Chauvac, Léa. *Les soeurs Nardal: À l'avant-garde de la cause noire*. Paris: Autrement, 2024. 185p. €21.00.

Mulira, Sanyu Ruth. "'We Were But Women': Print Culture, Anti-Colonialism, and Activism Amongst Black Women in the Francophone Diaspora, 1920–1976." Ph.D., New York University, 2023.

Murphy, Lacy. "Colliding Worlds: Visual Culture in France and Colonial Algeria (1918–1962)." Ph.D., Washington University in St. Louis, 2024.

Ofrath, Avner. *Colonial Algeria and the Politics of Citizenship*. London, UK: Bloomsbury Academic, 2023. 194p. $39.95.

Peterson, Terrence G. *Revolutionary Warfare: How the Algerian War Made Modern Counterinsurgency*. Ithaca: Cornell University Press, 2024. 240p. $46.95.

Pisanelli, Simona. *Slavery and Colonialism in the History of Economic Thought: The Cases of France and Great Britain*. Abingdon, Oxon: Routledge, 2025. 160p. $170.00.

Reste de Roca, Michèle-Alex. *Le gouverneur général F.-J. Reste: Afrique au cœur, 1879–1976*. Paris: Harmattan, 2024. 230p. €24.00.

Rosenstein, Brent Matthew. "The Face of the Nation: Drogmans and the French State, 1669–1880." Ph.D., State University of New York at Buffalo, 2023.

Rucker, Hayley R. "Thrown to the Sea: Capitalism and Belonging Aboard Ship on French Transoceanic Voyages, 1680–1793." Ph.D., University of California, Berkeley, 2023.

Sagnières, Hubert. *Daring French Explorations: Trailblazing Adventures around the World: 1714–1854*. Translated by Florence Brutton. Paris: Flammarion, 2024. 399p. $85.00.

Saïdi, Hédi. *La franc-maçonnerie dans la Tunisie coloniale: Saga de la loge l'Aurore du XXe siècle de Bizerte: 1900–1940*. Paris: Editions du Cygne, 2024. 187p. €20.00

Slyomovics, Susan. *Monuments Decolonized: Algeria's French Colonial Heritage*. Redwood City, CA: Stanford University Press, 2024. 330p. $35.00.

Stevens, Kate. *Gender, Violence and Criminal Justice in the Colonial Pacific, 1880–1920*. London: Bloomsbury Academic, 2023. 279p. $115.00

Tupinier, Jacques. *Main d'œuvre au Cameroun*. Edited by Jean-Pierre Le Crom. Paris: Classiques Garnier, 2024. 217p. €26.00.

Vermeren, Pierre, ed. *Qui a sauvé les harkis?: Témoignages et matériaux sur une histoire méconnue: Actes du colloque du 22 septembre 2022 aux Invalides à Paris "Hommage à ceux qui ont sauvé des Harkis" sous le haut patronage de Monsieur Emmanuel Macron, président de la République*. Paris: Riveneuve, 2024. 205p. €22.00.

Youssef, Ahmed. *Bonabarta: Napoléon, une passion arabe?* Paris: Passés/Composés, 2024. 149p. €17.00.

Translated Abstracts

MARGARET CARLYLE

**Les femmes et l'enseignement de l'anatomie dans la France des Lumières :
Un scalpel à double tranchant**

Cet article explore l'écart entre les objectifs ambitieux de l'anatomie au XVIIIe siècle en tant que discipline destinée à émanciper les femmes et les réalités complexes de leur engagement sur ce sujet, qui suscitaient à la fois l'admiration et la peur. Cet article soutient que les nouveaux supports utilisés dans l'enseignement médical, tels que les modèles en cire et les squelettes réels, étaient perçus à la fois comme des outils libérateurs et des sources d'anxiété morale. Les interactions des femmes avec ces modèles et les connaissances anatomiques qu'ils représentaient étaient souvent acceptées et même célébrées lorsqu'elles étaient présentées dans un contexte amateur et sous la supervision d'experts. Cependant, les modèles étaient perçus comme troublants lorsqu'ils étaient consommés de manière indépendante ou dans un contexte subversif. Cette tension a culminé à la fin du siècle avec une réaction violente contre les visites des deux sexes dans les musées de cire.

MOTS CLÉS   genre, pédagogie, modèles en cire, corps, moralité

YOTAM A. TSAL

Exporter la nature française : Les origines transnationales de la taxidermie

En 1819, une collection monumentale d'histoire naturelle a été transportée de Paris à Edimbourg. En 1820, le manuel de son propriétaire, Louis Dufrense, qui a été l'une des premières personnes à utiliser le terme « Taxidermie », a été traduit du français à l'anglais pour servir aux musées et aux voyageurs du monde anglophone. La recherche archivistique et les méthodes informatiques permettent de retracer le transport de la collection ainsi que la traduction et la réception de ce manuel. Cet épisode de l'exportation d'une version hautement raffinée, perfectionnée et réglementée du monde naturel a donné aux Français un avantage concurrentiel sur le marché international du début du XIXe siècle pour les connaissances et les spécimens d'histoire naturelle.

MOTS CLÉS   zoologie, histoire naturelle, musées, science internationale, Grande-Bretagne

ELIZABETH DELLA ZAZZERA

L'illumination à Paris sous la Restauration

Vers 1816, les industriels français ont commencé à débattre le gaz hydrogène, et leurs soucis rangeaient de l'impact de cette technologie sur l'industrie de l'huile de graines au rôle que le charbon devait jouer dans les plans d'industrialisation. En 1823, le débat entra dans la conscience publique quand la construction d'un gazomètre souleva des inquiétudes quant aux propriétés explosives du gaz. En examinant la controverse sur l'éclairage par le gaz dans le contexte culturel de la Restauration et en explorant ses parallèles avec les conflits contigus, cet article montre comment l'éclairage est devenu plus qu'une question de technologie, d'illumination, ou d'industrialisation. Le débat sur l'éclairage par le gaz ne reflétait pas simplement une opposition de libéraux pro-technologie et de réactionnaires anti-modernité : le gaz hydrogène a pris de nombreuses significations changeantes. Ainsi, s'opposer au gaz pouvait être considéré comme prudent et rationnel et le soutenir comme antipatriotique et imprudent. Le gaz pouvait être vu comme un affront esthétique ou une ressource spectaculaire ; il pouvait être imaginé comme une menace à la souveraineté française ou comme un instrument de l'état.

MOTS CLÉS gaz hydrogène, romantisme, classicisme, technologie, industrialisation

VOLNY FAGES, JÉRÔME LAMY, and FLORIAN MATHIEU

Louise Michel and the Knowledge of Exile: The Socio-Epistemic Crossings of the "People's Minerva" in the Penal Colony of New Caledonia

Louise Michel, condemned to a penal colony after the Commune, was sent to New Caledonia from 1873 to 1880. During this period, she produced and transmitted knowledge in three fields of scholarly investigation. First, she practiced a natural science combining academic knowledge with sensitive formulations of knowledge. Second, she worked on an ethnology of the Kanak populations, torn between imperialist reflexes and emancipatory empathy. Finally, she was committed to the transmission of knowledge. In these three scholarly sectors, Louise Michel combined a political approach to knowledge (including analogies between the natural world and human societies) and an academic outlook (acknowledging the importance of scientific institutions). Without ever completely adhering to the expectations of each academic, militant, and penitentiary social space, her ability to traverse them while maintaining her own political and scholarly requirements indicates an original positioning between the knowledge of the margins and that of the academic centers.

KEYWORDS penal colonies, amateurs, natural science, ethnology, Commune

LOUISE THIROUX

Women's Toilets Take Over the Streets: An Architectural and Engineering Study of Parisian Water Closets (1872–1900)

This article traces the history of a little-known type of Parisian street furnishing: les chalets de nécessité, or public water closets. Available in two main forms, these places of comfort flourished in the capital between 1872 and 1900. This sociocultural history of technology and architecture highlights the various ways in which street furnishings interacted with the hygienic discourse of the time, forging a special place for what we call the female pissing body at the end of the century.

KEYWORDS public toilets, Paris, technology, urban history, women's history

ZOHAR SAPIR DVIR

Automobiles, entrepreneurs et Empire colonial français : L'auto-tourisme en Afrique du Nord entre les deux guerres

A la fin de la Première Guerre mondiale, les dirigeants d'entreprises ont commencé à promouvoir l'automobile comme le moyen idéal pour visiter l'Afrique du Nord française. De grandes entreprises comme Dunlop, Michelin, la Compagnie générale transatlantique et Citroën ont mené des initiatives pour promouvoir les voyages en voiture au Maroc, en Algérie et en Tunisie. Ces efforts s'alignaient sur l'intérêt croissant de l'administration coloniale française pour la technologie, et en particulier pour une mobilité motorisée améliorée, comme moyen de développer, maintenir et gouverner son empire nord-africain. Cet article analyse le croisement entre la technologie et le colonialisme en Afrique du Nord française durant l'entre-deux-guerres à travers le prisme du tourisme automobile. En remettant en question la division entre politique, technologie et loisir, il examine les usages idéologiques et les effets matériels de l'auto-tourisme du point de vue de la mobilité, soulignant le rôle de la mobilité coloniale liée aux loisirs dans la formation des politiques impériales et des pratiques coloniales françaises de l'entre-deux-guerres.

MOTS CLÉS   tourisme colonial, guides touristiques, mobilité, technologie, histoire de l'entreprise

Eugen Weber Book Prize

The Department of History at UCLA encourages submissions for the 2026 Eugen Weber Book Prize in French History. A prize for the best book in modern French history (post 1815) over the previous four years, this award is named for eminent French historian Eugen Weber (1925-2007). Professor Weber served on the History faculty at UCLA from 1956 until 1993 and was renowned as a teacher and scholar for being able to bring the French and European past to life.

The Eugen Weber Book Prize in French History brings a cash award of $15,000 and the winner will be announced at the American Historical Association annual meeting in January 2026. The author will be invited to visit UCLA to speak about his or her work and receive the prize during the spring of 2026.

Books eligible for the 2026 prize are those written in English or French and published in 2023 or 2024.

The deadline for submissions is June 1, 2025. Submission information is available at https://history.ucla.edu/eugen-weber-book-prize.

The prize was most recently awarded in 2024 to Owen White for The Blood of the Colony: Wine and the Rise and Fall of Algeria (Harvard University Press, 2021) and Marc André for Une prison pour mémoire: Montluc, de 1944 à nos jours (ENS Éditions), in 2022 to Judith G. Coffin for Sex, Love, and Letters: Writing Simone de Beauvoir (Cornell University Press, 2020) and in 2020 to Christine Haynes, for Our Friends the Enemies: The Occupation of France After Napoleon (Harvard University Press, 2018)

For more information, visit http://history.ucla.edu.

UCLA Meyer and Renee Luskin **Department of History**

Keep up to date on new scholarship

Issue alerts are a great way to stay current on all the cutting-edge scholarship from your favorite Duke University Press journals. This free service delivers tables of contents directly to your inbox, informing you of the latest groundbreaking work as soon as it is published.

To sign up for issue alerts:

1. Visit **dukeu.press/register** and register for an account. You do not need to provide a customer number.

2. After registering, visit **dukeu.press/alerts**.

3. Go to "Latest Issue Alerts" and click on "Add Alerts."

4. Select as many publications as you would like from the pop-up window and click "Add Alerts."

read.dukeupress.edu/journals    **DUKE** UNIVERSITY PRESS